Potent Compounds from Seagrasses Against Deadly Vectors

by

D. Monisha

CONTENTS

GLOSSARY OF SYMBOLS AND ABBREVATIONS

±	Plus or minus
°	Degree Celsius
μg	Microgram
μg/cm^{-2}	Microgram centimetre power -2
μg/L	Microgram per Litre
μg/ml	Microgram per milllitre
μl or μL	Microliter
μm	Millimicron
μM	Micrometre
3D	Three dimentional
Ace1	Acetylcholinesterase gene1
AChE	Acetylcholinesterase enzyme
Ag	Silver
AgNO$_3$	Silver Nitrate
AgNPs	Silver nanoparticles
Atm/psi	Atmospheric pressure
ATR	Attenuated total reflectance
BHC	Benzene Hexachloride
BLAST	Basic Local Alignment search tool
CDCl$_3$	Deuterated chloroform
Cm	Centimetre
cm^{-1}	Centimetre power -1
Df	Film thickness
DMSO	Dimethyl methoxysulphur
DPX	Dibutylphthalate polystyrene
EC$_{50}$	Half Maximum effective concentration
EC$_{50}$	The dose concentration producing 50% response
EDX	Energy Dispersive X-ray spectroscopy
ev	Electronic charge
FASTA	Fast alignment Search tool
FE-SEM	Field Emission scanning electron microscopy
FTIR	Fourier Transform Infra-red spectroscopy
G	Gram
GC-MS	Gas Chromatography Mass Spectroscopy
GPS	Ground Positioning system
H$_2$SO$_4$	Sulphuric acid
HCl	Hydrochoric acid
K cal	Kilo calorie
LC$_{50}$	Concentration of Compound which kills 50 Percent test organisms
LC$_{90}$	Concentration of Compound which kills 90 Percent test organisms

LCMS	Liquid chromatography Mass Spectrometry
LD_{50}	Lethal Dose concentration to kill 50% of tested organism
LD_{90}	Lethal Dose concentration to kill 90% of tested organism
M^+	Molecular ion
m/z	Mass to the charge ratio
mg	Milligram
mg/L	Milli gram per litre
mg/ml	Milli gram per millilitre
MHz	Mega Hertz
Min	Minutes
ml	Milliliter
mm	Millimetre
Mol	Mole
mp	Melting point
N	Normality
NaoH	Sodium hydroxide
NIST	National Institute of standards and technology
Nm	Nano metre
NMR	Nuclear Magnetic Resonance
OD	Optical Density
Pd	Pelladium ions
PDB	Protein Data Bank
ppm	Parts per million
Rh	Relative humidity
RT	Room temperature/ Retention time
SEM	Scanning Electron Microscopy
Sp.	Species
TEM	Transmission Electron Microscopy
$Ti\,O_2$	Titanium ions
TLC	Thin Layer Chromatography
UV-Vis	Ultra violet Visible
Viz.,	As follows
w/v	Weight per volume
XRD	X-ray diffraction
α and β	Alpha and beta
Δ	Delta

1. INTRODUCTION

Mosquito Problem:

Mosquitoes are flying insects that have a greater impact on public health. They belong to the order Diptera and family of Culicidae. Mosquitoes play a predominant role in transmitting dreadful diseases such as malaria, filariasis, japanese encephalitis, dengue, chikungunya, yellow fever and zika fever for that reason they carry pathogenic viruses and parasites (Hafeez *et al.*, 2011; Samidurai *et al.*, 2009; Elumalai *et al.*, 2013). Mosquitoes are broadly disbursed in tropical and subtropical zones of Asia, Africa and some parts of America (Fradin and Day, 2002; McHugh, 1994; Hales *et al.*, 2002).

There were about 3000 species of mosquitoes that have been reported to circulate diseases in the globe now and they belong to three Genus: *Anopheles*, *Culex* and *Aedes* which are responsible for human bites (Rozendaal, 1997; Knight and stone, 1977). In particular, female mosquitoes have an extended piercing needle like mouth part called proboscis which injects the saliva and they suck the desired blood meal to get protein which is necessary for egg production. The saliva contains toxins, harmful allergic chemicals which comprise the skin and causes red dark spots, irritation, small blisters and bumps that leads to fever and other infections (Jayne leonard, 2018).

According to the latest declaration of World Health Organisation (WHO, 2020), it was assessed around 229 million people are infected and 4,09,000 people have died due to the cause of malaria and particularly the children below five years are considered more vulnerable and 2,74,000 (64%) was died worldwide in 2019. This was released by World malaria reports of 2020, the malarial cases are gradually increasing over the years from 228 million cases in 2018 to 229 million cases in 2019. This shows that still more actions need to be taken by the WHO in the years to come.

Aedes aegypti and *Aedes albopictus* mosquitoes are the most widespread day-biting dengue vectors, they breed in the clean water bodies. Both the *Aedes* Sp. transmits infectious arbovirus of the family flavivirus to human beings, which are responsible for causing dreadful diseases such as dengue, yellow fever and chikungunya (Samidurai *et al.*, 2009; Tolle, 2009). Another common vector, *Culex quinquefasciatus* is the major night-biting filarial vector that transmits parasites to the human beings. They breed only in the dirty polluted waters and drainages which causes encephalitis and lymphatic filariasis (Vijayakumar and Amirthanathan, 2014). Nevertheless, these three mosquito vectors (*Aedes aegypti*, *Aedes albopictus* and *Culex quinquefasciatus*) have their own distinctive pattern of breeding habits and it is effective to control them in their immature stages (larval stage) to achieve maximum control of their population in the breeding sites.

Synthetic insecticides in mosquito control:

Application of adulticide is not an effective control measure, because adult population of mosquitoes are free-living and not restricted in a particular place (Autran *et al.*, 2009; Severini *et al.*, 1993). In the present circumstances, the mosquito larvae are largely controlled by the regular insecticides of organosphosphates such as temephos, fenthion, biological insecticides including *Bacillus* formulations and growth regulators like, diflurobenzuron and methoprene (Lin *et al.*, 2010). Pyrethrin is a chemical compound used in the mosquito repellents (Sa Ra *et al.*, 2009). The repeated use of the chemical insecticides will damage the natural food chain that creates environmental pollution problems, hazardous to human beings and destroy other target organisms (Suwanbamrung *et al.*, 2010). This leads to demand, the need of finding sustainable organic insecticides which are eco-friendly and effective in controlling the organisms (Girdhar *et al.*, 1988).

Natural Insecticides:

Plant based biocides are finding the major role in the insecticides that also replace the fossil based synthetic insecticides in the recent years due to its biodegradability, eco-friendly, less pollution and less harmful to non-target organisms (Nasar *et al.*, 2009; Yusniawati *et al*, 2017). From this standpoint, many plant extracts produce secondary metabolites and their derivatives have shown insecticidal growth inhibition, larvicidal and repellent efficiency. This is why, they can be used as commercial insecticides to screen and kill the mosquito larvae (Girdhar *et al.*, 1988; Thangam and Kathiresan, 1990, 1994; Samidurai *et al.*, 2006).

In the past few years, researchers across the world are now focusing their research works on marine organisms to obtain eco-friendly insecticides. Marine algae, seaweeds and other marine natural products contain novel and valuable secondary metabolites with anti-viral and anti-fungal properties (Maye *et al.*, 2009) and insecticidal properties (Thangam and Kathiresan, 1988; Kabaru and Gichia, 2001). Based on the good results shown these natural eco-friendly insecticides are considered as an alternative sources of vector control agents (Venkateswara Rao *et al*, 2008). Despite everything, there are various types of medicines and complex compounds yet to be produced from marine organisms that can open doors for many therapeutic treatments (Nasar *et al*., 2009; Prakash *et al*., 2007).

Seagrass as potential natural sources of bioactive compounds:

Several studies have been carried out against mosquito larvicidal activity on microalgae and macroalgae in the past years but only minimal work was carried out on seagrasses which is beneficial because of their lifecycle, huge availability in marine diversity and presence of secondary metabolites (Montano *et al*., 1999; Puglisi, 2007). Seagrasses are marine flowering plants (Angiosperms) tethered with the help of mud and sand in the deep bottom of aquatic biodiversity below 1-10 feet, based upon species adaptability (Den Hartog and John Kuo, 2006). Donald and Maryke, 1997 reported four families Cymodoceaceae, Hydrocharitaceae, Zosteraceae and Posidoniceae, 12 genera and 60 species of seagrasses. Seagrasses are abundantly present in tidal and subtidal areas than in coastal regions and its morphological characteristics such as leaves, rhizome, stem and reproductive organs are found prominent and extraordinary (Pushpa bharathi and vanitha, 2017). Seagrasses helps energy production and shore stabilisation by absorbing toxic elements, providing nutrient supplements to other marine organisms. Besides this, the seagrass acts as an antifouling agent and environmental indicator as well. (Fortes, 1989; Malea *et al*., 1995; Caccia *et al*., 2003; Ferrat *et al*., 2003).

In accordance with Hemminga and Duarte (2000), the coastal populations use the biomass of seagrasses as food and also used in the treatment of fever, skin diseases, muscle pains, wounds, stomach problems, a remedy against stings of different kinds (De la Torre-casto and Ronnback, 2004). Seagrasses have all the pharmacological efficacies such as antimicrobial, anti-mosquito larvicide, antioxidant, antipyretic and antitumor attributable to the presence of certain secondary metabolites such as flavonoid, terpenoids, phenols, alcohol and steroids *etc.* (Ravi kumar *et al*., 2005; Subhashini *et al*., 2013; Papenbrock, 2012). Kannan *et al*., 2013 examined phytochemical activity, coumaric acid analysis and antioxidant analysis of some seagrasses which were collected from Gulf of Mannar, biosphere reserve, Tamil Nadu India.

Mahyoub *et al.*, 2017 stated that Silver nanoparticles (*AgNPs) Halodule uninervis* has an efficient reducing and capping agent. Amudha *et al.*, 2019 reported that Ag (silver), Pd (palladium ions) TiO$_2$ (Titanium di oxide ions) synthesized from *Cymodocea serrulata* was tested against three different larvae *Aedes aegypti, Culex quinquefasciatus* and *Anopheles stephensi*. Abdhul *et al.*, 2015 explored the antioxidant property of Silvernanoparticles (*AgNPs)* from *Cymodocea serrulata* found to be efficient against *HeLa* cells and *vero* cells. Yuvaraj *et al.*, 2012 stated methanolic extracts of *Halophila* Sp. to have strong scavenging activity, efficient antibacterial activity and promising anti-inflammatory activity.

Kannan *et al.*, 2010 evaluated the antibacterial activity of hexane, chloroform and methanolic extracts of three seagrasses such as Halodule uninervis, Halophila ovalis and Cymodocea serrulata was against seventeen human pathogens. Danaraj *et al.*, 2016 evaluated the anti-oxidant activity from leaves and rhizome of 50% aqueous methanolic extracts of *Halophila uninervis, Syringodium isoetifolium, Cymodoceae roundata, Enhalus acoroides, Thalassia hemprichii* and *Halophilia ovalis*. Kannan *et al.*, 2013 evaluated antibacterial, cytotoxic assay and haemolytic activity of aqueous methanolic extracts of *Enhalus acoroides, Thalassia hemprichii, Halodule pinnifolia, Syringodium isoetifolium, Cymodoceae serrulata* and *Cymodoceae roundata*. Palaniappan *et al.*, 2014 reported the cytotoxic activity of *AgNPs Cymodocea serrulata* against human lung cancer *A549* cell lines. Perez *et al.*, 2018 exhibited anti-microbial property of *Cymodocea rotundata*. Nanthakumar *et al.*, 2013 explored antioxidant property on *Cymodocea serrulata*. Sudharsan *et al.*, 2012 exhibited heavy metal absorption from *Syringodium isoetifolium* and *Cymodocea serrulata*. Grennan, 2011 reported the same from *Halodule uninervis* and *Halophila pinifolia*.

Padmanaban *et al.*, 2018 reported *AgNPs* from *Cymodocea serrulata* has examined the antibacterial efficiency against prawn pathogen *Vibrio parahaemolyticus* and its combative effect on shrimp *Pandenaeus monodon*. *Cymodoceae serrulata* used as tranquilizer for infants and also acts as anticancerous agent against *HeLa* Cells (Hardoko *et al.*, 2016). Girija *et al.*, 2013, exhibited anti-coagulant activity on account of the presence of secondary metabolite phenol. Rowley *et al.*, 2002 screened *Thalassia testudinum* (Caribbean seagrass), isolated HIV integrase inhibitors, thalassiolins A-C and provoked its antiviral behaviour. Kannan *et al.*, 2012 examined antibacterial activity of *Enhalus acoroides, Thalassia hemprichii, Halodule pinnifolia,*

Syringodium isoetifolium, Cymodoceae serrulata, Cymodoceae roundata against urinary tract pathogens. Hua *et al.*, 2006 investigated the methanolic extracts of *Zostera japonica* against anti-inflammatory activity. Gokce and Haznedaroglu, 2008 screened antioxidant, antidiabetic and vasoprotective activities from seagrass *Posidonia oceanica*. Qi *et al.*, 2008 studied the potentiality of *Enhalus acoroides* against antibacterial, anti-viral, anti feedant properties whilst, Alino *et al.*, 1991 screened the aphrodisiac and contraceptive properties from *Enhalus* seeds. Kannan *et al.*, 2012 explored the antioxidant activity of *Halophilia ovalis, Halophilia ovata, Halophilia stipulaceae, Thalassia hemprichii, Syringodium isoetifolium* and *Halodule pinnifolia*, collected from the intertidal region at Mandapam coast of South India.

Angel and Uma Maheswari *et al.*, 2015 stated that *Syringodium isoetifolium* exhibited anti-fungal activity against six common fungal pathogens. Kannan *et al.*, 2010 studied *in-vitro* anti-oxidant activity from ethanolic extracts of *Enhalus acoroides*. Kannan *et al.*, 2010 examined the thin layer chromatography and antioxidant analysis of aqueous methanolic extracts of four seagrasses collected from Gulf of Mannar, South India. Ravi kumar *et al.*, 2011 exhibited antibacterial activity from root extracts of *Cymodoceae serrulata* against poultry pathogens.

The Compound thiopolyketone was isolated from *Cymodoceae serrulata* against fourteen clinical bacterial pathogens (Gnanambal *et al.*, 2015). The presence of tannin content in *Halodule pinifolia*, exhibited antibacterial activity against gram positive and gram negative bacteria (Supriadi *et al.*, 2016). Antibacterial compound biarane diterpene (Kontiza *et al.*, 2008), cymodienol, cymodiene (Kontiza, 2005); 3-keto steroids (Kontiza *et al.*, 2006) were isolated from *Cymodocea nodosa*. Zosteric acid (Achamlale, 2009a); rosmarinic acid (Achamlale, 2009b) was isolated from *Zostera noltii* and *Zostera marina*. Syphonosid was isolated from *Halophila stipulaceae* (Carbone *et al.*, 2008). Chicoric acid from *Posidonia aceanica* (Haznedaroglu and Zeybek, 2007) and *Syringodium filiforme* (Nuissier *et al.*, 2008). Ravi kumar *et al.*, 2011 reported antibacterial efficacy from root extracts of *Cymodoceae serrulata* against bacterial fish pathogens. Mayavu *et al.*, 2009 reported antibacterial activity of seagrasses against bio-film forming bacteria.

Mosquito larvicidal activity of seagrasses:

The different seagrasses were screened against mosquito larvae of *Aedes aegypti* (Ali *et al.*, 2012; Ali *et al.*, 2013; Vijayakumar *et al* 2014; Mahyoub *et al.*, 2016; Yusniawati *et al.*, 2017; Purnomo *et al.*, 2017), different seagrasses against *Culex quinquefasciatus* was studied by Vijayakumar and Amirthanathan, 2014; Devi *et al.*, 1997, seagrasses against *Culex pipens* was studied by Khattab *et al.*, 2012. Nonetheless, these research works were carried out only in crude extract with few polar solvents but, there is no better understanding about increasing order of solvent selection, active compound isolation and mechanism of larval mortality but, the current research aims to carry out a complete research on larvicidal activity of seagrasses extracted with increasing polarity of three different solvents, active compound isolation, exploration *Viz.*, bio guided fractionation, spectroscopic studies and its binding efficacy through *in silico* docking analysis.

2. OBJECTIVES

---

1. Collection and identification of selected seagrasses from Mandapam coastal regions, Ramanathapuram, Tamil Nadu, India

2. Preliminary phytochemical screening of the seagrasses using different solvents

3. Collection, isolation, identification, and culturing deadly vectors of two different species of mosquito larvae *Aedes aegypti* and *Culex quinquefasciatus*

4. Analyse the larvicidal potential of crude solvent extracts of selected seagrasses against third instar larvae of *Culex quinquefasciatus* and *Aedes aegypti*

5. Screening and isolation of the potent larvicidal compound by bioassay guided fraction method

6. Separation and chemical characterization of the potent larvicidal compound using TLC, UV-Vis spectroscopy, FTIR, ^{1}H NMR and GC-MS

7. Examination of histopathological study of mosquito larvae against the potent larvicidal compound isolated from seagrasses

8. Analyse the non-target effect of guppy fish (*Poecilia reticulata*) and mosquito fish (*Gambusia affinis*) against the potent larvicidal compound

9. To perform *in silico* molecular docking analysis of homology model of *ace1* receptor of mosquito larvae against the potent larvicidal compound

3. REVIEW OF LITERATURE

Vector borne diseases:

Vector-borne diseases threaten most of the species globally and it was proclaimed that the issue is still worst in the place of poor sanitation, lack of clean drinking water and poverty line (WHO, 2014; 2016). To control the vector-based diseases in the global market today, the scientific community tries to monitor applications of chemical insecticides that impact our natural environmental conditions. For example, the insecticide called DDT (Dichloro diphenyl trichloro ethane) spoils the soil fertility and affects the aquatic and birds lifestyle too (Casida and Quistad, 1998). Likewise, there are other insecticides that incorporate carbamates and organophosphates that are harmful to the other biological organisms such as livestock, pollinating birds, wild animals, aquatic life and human beings (Jeyaratnam, 1990; Kwong, 2002; Botha *et al.*, 2015; Verster *et al.*, 2004; Eskenazi *et al.*, 1999). In recent years the world has started looking at natural compounds which leads to progress in many studies in drug development and synthetic insecticides which proved to be safe in the results (Abou-Elnaga *et al.*, 2011; Adams, 2007). In subsequent days, marine natural products started to attract its unique chemical structures and functions that could control mosquito larvae with less cost (Selvin and Lipton, 2004). Nevertheless, based on earlier investigations, it shows that quite a little research work has been performed on larvicidal studies by analysing seagrasses and more research works yet to be explored in the coming days.

Mosquito vector diseases:

Malaria is caused by *Anopheles* mosquito (Torrades, 2001), Yellow fever is caused by *Aedes aegypti* (Bisset *et al.*, 2009), *Haemogogus jantinomys* and *Sabethes choropterus* (GVCET, 2010), Dengue is caused by *Aedes aegypti* and *Aedes albopictus* (Barnett, 2007), Venezuelan equine encephalis is caused by *Culex vomerifer*, *Culex pedroi*, *Culex adamesi* and *Aedes taeniorhynchus,* (Brown and Condreay, 2003), Japanese encephalitis is caused by *Culex tritaeniorhynchus, Culex annulus, Culex annulirostris, Culex vishnui, Culex fuscocephala* and *Culex gelidus* (Hammerer, 2007). Lymphatic filariasis is caused by *Anopheles, Aedes, Culex* and *Mansonia* Sp. Rift valley fever (CFSPH, 2006) and West nile fever is caused by *Aedes* and *Culex* (EFSA, 2013; Granwehr *et al.,* 2004), Chikungunya fever is caused by *Aedes aegypti* and *Aedes albopictus* (Porta, 2012) Saint louis encephalitis is caused by *Culex, Aedes, Mansonia* and *Sabethes* genera (Rodrigues *et al.*, 2010), Eastern equine encephalis is caused by *Coquillettidia perturbans, Culiseta melanura, Ochlerotatus canadensis, Aedes vexans* and *Culex* (NYSDOH, 2012), Western equine encephalitis is caused by *Culex* and *Culiseta* genera, *Culex tarsalis* (Ray, 2004; OIE, 2013; Mahmood *et al.*, 2004), La crosse encephalitis is caused by *Aedes* genus and *Aedes triseriatus*, Zika fever is caused by *Aedes africanus* and *Aedes aegypti* (CDC, 2014).

Life cycle of *Culex* and *Aedes* species:

Depending on environmental conditions such as temperature, light and relative humidity, life cycle of *Culex* and *Aedes* mosquito persists distinctive life cycle though it has diverse morphology. It is classified into two stages (Fig 1a; Fig 1b)

1. Breeding stage/Immature stage
2. Flying stage

Breeding stage or immature stages will be spent in water majorly, consists of egg, larvae and pupa. *Aedes* lay prominent isolated eggs on the outer surface of water, hatches 150-200 larvae. *Culex* eggs lay the eggs above the water surface, they are glued and stacked in the form of miniature rafts, hatches 100 - 400 larvae within 24 - 48 hours, their oviposition may persist from three months to two years. The larval stage is classified in to four instars with respect to increase in size of body (first, second, third and fourth). The larvae breathe atmospheric oxygen with the help of siphon. After fourth instar stage, it metamorphosis into pupa, non-feeding or resting stage, exhibit tumblery motion, the process takes place between 4-6 days, an intermediate period between breeding and flying stage.

Flying stage:

It takes 24 to 48 hours to stretch their wings to become adult (complete development). They require carbohydrate energy (sucrose solution, nectar, plant exudates) to sustain their life to fly, mate and to select specific host to obtain desired blood (protein) for its egg production. The overall lifecycle takes place between 8-10 days (Tennyson *et al.*, 2007; Mike service, 2008; Reinert *et al*, 2004, Clements, 2000).

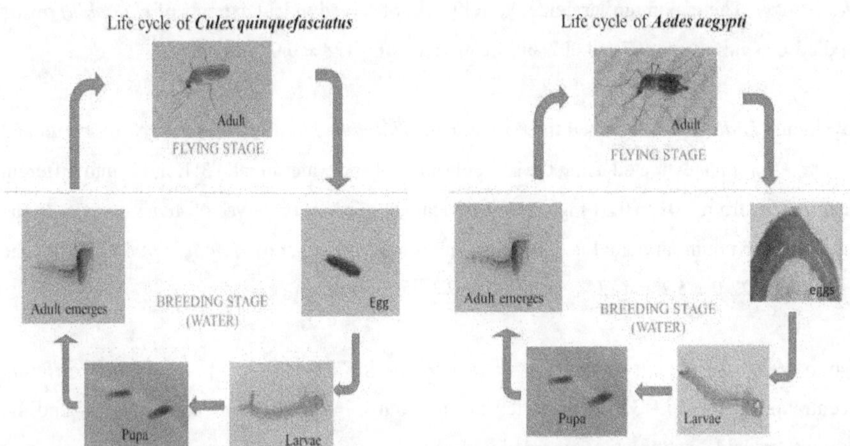

Fig.1a. Life cycle of *Culex quinquefasciatus*; Fig.1b. Life cycle of *Aedes aegypti*

Mosquito larvicidal activity:

In accordance to the research criteria and feasibility, first instar to fourth instar larvae was conceivably selected. Countably, twenty-five larvae were introduced into a beaker containing 250 ml of water mixed with different concentrations of solvent extracts from 62.5-500 ppm including its corresponding replicates. The mortality was tested after 24 hours, exceeding that the larvae do not show mobility after shaking with needle, the larvae are considered as dead (WHO, 2005), the percentage of mortality assessed through Abbott's formula, 1925.

Larvicidal activity of seagrasses:

Ali *et al.*, 2012, deserted three seagrasses such as *Syringodium isoetifolium*, *Cymodocea serrulata* and *Halophila beccarii*, extracted with ethanol:water (3:1) solvents, prepared into different concentration ranging from 0.01 - 0.1 mg, treated against fourth instar larvae of *Aedes aegypti*, the maximum larvicidal activity was reported in the root extract of *Syringodium isoetifolium* with the LC_{50} of 0.060 ± 0.005 and LC_{90} value of 0.909 µg/ml.

Ali *et al.*, 2013, screened four different seagrasses *Halophila ovalis*, *Enhalus acoroides*, *Thalassia hemprichii* and *Halodule pinifolia*, extracted with ethanol: water (3:1) solvents, made into various concentrations from 0.01 - 0.1 mg and tested against fourth instar larvae of

Aedes aegypti. The maximum larvicidal activity was observed in leaf extracts of *Halophila ovalis* with the LC_{50} value of 0.067 ± 0.007 and LC_{90} value of 0.128 ± 0.025 µg/ml.

Vijayakumar *et al.*, 2014, screened three seagrasses *Halodule pinifolia, Cymodocea serrulata* and *Thalasia testudinum* extracted using the solvents ethanol and water in ratio 3:1, made into different concentrations from 0.01 to 0.01 mg, examined against fourth instar larvae of *Aedes aegypti.* In the results, the maximum larvicidal activity was found in root extract of *Halodule pinifolia* with the LC_{50} value of 22.0 ± 5.2 and LC_{90} value of 54.2 ± 2.5 µg/ml.

Mayoub *et al.*, 2016, extracted *Thalasia hemprichii* in 70% ethanol, prepared into different concentration from 150 - 300 ppm, tested against fourth instar larvae of *Aedes aegypti* and the maximum larvicidal activity was recorded with the LC_{50} value of 201.7 ppm.

Yusniawati *et al.*, 2017, screened *Thalassia hemprichii* collected from two different places Prapat Agung and Menjagam Island, extracted with 1% methanolic solvent, screened against third instar larvae of *Aedes aegypti* and the final evaluations proved that 100% mortality was observed in leaf extracts of *Thalassia hemprichii* (Prapat Agung) and root extracts of *Thalassia hemprichii* (Menjagam Island) after 24 hours of treatment.

Vijayakumar and Amirthanathan, 2014 screened *Halodule pinnifolia, Cymodoceae serrulata* and *Thalassia testudinum* extracted using ethanol: water (3:1) solvents, fabricated into different concentrations 0.01- 0.1 mg against fourth instar larvae of *Culex quinquefasciatus.* The maximum larvicidal activity was found in root extracts of *Thalassia testudinum* with the LC_{50} value of 0.061 $\pm$ 0.005 and LC_{90} value of 0.912 µg/ml.

Purnomo *et al.*, 2017, examined the efficacy of leaf, rhizome and root extracts of *Thalassia hemprichii* produced using methanol: water (3:1), tested against third instar larvae of *Aedes aegypti* and the mortality was observed in 12, 24 and 48 hours, from the results100% lethality was found in leaf, root extract of *Thalassia hempricii* and 56% of lethality was reported in rhizome extract.

Devi *et al.*, 1997, screened petroleum ether and chloroform soluble fractions of methanolic extracts of seagrasses such as *Halophila ovalis* and *Syringodium isoetifolium*, tested against third and fourth instar larvae of *Culex quinquefasciatus*. From the bioassay, it was found *Syringodium isoetifolium* was ineffective till 450 mg/L in chloroform fractions and ineffective till 2000 mg/L in petroleum ether fractions, both the fractions of *Halophila ovalis* were ineffective till 2000 mg/L.

Kattab *et al.*, 2012, presented *Halodule uninervis* and *Halophila stipulacea* extricated using ethyl acetate and made into different concentrations, tested against third instar larvae of *Culex pipens* and it was observed that *Halophila uninervis* was more competent than *Halophila stipulacea*.

Srinivasan *et al.*, 2018, cross-examined the polar solvent extracts of *Cymodocea serrulata*, processed into different concentrations of 0.625, 1.25, 2.5 and 10 µg/ml, tested against late third instar and early fourth instar larvae of *Aedes aegypti*. The mortality was observed with the LC_{50} and LC_{90} value of 2.34 µg/ml and 15.49 µg/ml.

Review on Epiphytes of seagrasses:

Epiphytes are sessile or peri planktonic organisms that include red algae, cyanobacteria, bacteria, fungi, sponges, bryozoans, protozoa dwells on seagrasses, the older parts of plant are found accord with more epiphytes whereas new plants are less influenced, the habitat and distribution depends on the factors such as light, climatic changes, temperature, nutrients, predator interaction and water movement rely on. The life span and development of the epiphytic organisms depends on species, habitats and seasonal changes, they act as a primary producer in the marine ecosystem, so it can contribute more to the food web and also help in nutrient cycling, certain epiphytic bacteria fix the atmospheric nitrogen in the rhizosphere and some epiphytic cyanobacteria fix nitrogen in the leaves of seagrasses. It also functions in the formation of calcium carbonate algal sediments so they are often called as "sediment formers". After a certain period of time, these epiphytes become excessive due to over loading of nutrients on them sometimes may die when it could not receive adequate sunlight (Larkum *et al.*, 2006).

Larvicidal activity of micro algae:

Vincent *et al.*, 2011, examined five different acetone extracts of *Oscillatoria* Sp. The extract was treated with different metals such as copper, zinc, cadmium, chromium and control (untreated metals), prepared into five different concentrations of 50, 100, 150, 200 and 250 mg/L. They were tested against early fourth instar larvae of *Culex quinquefasiatus*. The maximum effect was observed in a *Oscillatoria* treated in zinc with LC_{50} value of 73.75 mg/L and LC_{90} value of 151.78 mg/L after 24 hours.

Ragunatha Rao, 1999, reported that methanolic extracts of *Westiellopsis* were taken in five different concentrations of 62.5, 31.2, 15.6, 7.8, 3.9 and 1.95 mg/L assessed against fourth instar larvae of *Aedes aegypti, Anopheles stephensi, Culex quinquefasiatus* (madurai), *Culex quinquefasiatus* (kochi), *Culex tritaeniorhynchus* and the maximum mortality was observed against *Aedes aegypti* with the LC_{50} value of 3.95 mg/L after 48 hours.

Rohani *et al.*, 2001, isolated aqueous extracts of four different chlorophytes such as *Chlorella vulgaris, Scenedesmus quadricauda, Chlorococcum* Sp. and *Ankistrodesmus convolutes*, collected at stationary phase, diluted further to obtain concentration (OD 620 nm =0.6). They were tested against 25 second instar larvae of *Aedes aegypti*, after 6 days of observation, the mortality rate was found to be 100 % in *Chlorella vulgaris,* 88 % mortality in *Chlorococcum* Sp. and 84 % mortality in *Scenedesmes quadricauda* and same type of study was again investigated by Rohani *et al.*, in 2004, screened ten chlorophytes *Scenedesmus* UMACC 010, *Ankistrodesmus convolutus* UMACC 101, *Chlorella* UMACC 184, *Chlorella* UMACC 185, *Chlorella* UMACC 187, *Chlorella* UMACC 193, *Chloroccocum* UMACC 213, *Chlorella* UMACC 217, *Chloroccocum* UMACC 218 and *Scenedesmus quadricauda* UMACC 220, were tested against 25 second instar larvae of *Aedes aegypti* and the seven species *Scenedesmus* UMACC 010, *Chlorella* UMACC 184, *Chlorella* UMACC 185, *Chlorella* UMACC 187, *Chlorella* UMACC 193 and *Chlorella* UMACC 217 shown 100 % mortality after 24 hours.

Rey *et al.*, 2009, choosed following marine and fresh water microalgae species *Scrippsiella* Sp. *Nitzschia kuetzingiana, Akashiwo sanguinea, Skeletonema costatum, Chlorella pyrenoidosa, Chlorella pyrenoidosa, Thalassiosira weissflogii, Nitzschia palea, Melosira lineata, Pandorina morum* and *Prorocentrum micans* were collected from marine water, 4 species from fresh river water of florida, *Scenedesmus quadricauda UTEX76, Chlorella vulgaris UTEX259, Chlorella vulgaris UTEX2714, Chlorella vulgaris UTEX1809,* 2 species of *Microcystis aeruginosa* (a) contains toxins and non-toxins, *Microcystis aeruginosa* (b) contains toxins only, were collected from Syracuse university, cultured the biomass and harvested in log and stationary phase, filtered, concentrated and treated against *Aedes aegypti*. The highest mortality was observed in marine water algae *Akashiwo sanguinea* and fresh water algae *Microcystis aeruginosa* (b)-toxins only.

Khasadan *et al.*, in 2003, disclosed that the combination of genes from *Bacillus thuringiensis* sub sp. *israelensis* has specific genes such as *cry4A, cry11Aa, cyt1Aa* and *p20* (regulatory gene) introduced into the *Anabaena* PCC 7120 strain that has two promoters *PpsbA* and P_{AI}, responsible to control expression. As a result of gene combination, clone pRVE4-ADRC was developed, tested against the fourth instar larvae of *Aedes aegypti*. From the examination, clone has been reported to have 5-fold toxicity to kill the mosquito larvae compared to Wu *et al.*, who reported the clone to have 2.5-fold efficacy without the *cyt1Aa* gene in 1997.

Sigamani *et al.*, 2020, conducted a study on different solvent extracts of hexane, chloroform, methanol and ethanol of *Chlorella* Sp. tested against third instar larvae of *Aedes aegypti*. After the conduct of assay, the highest mortality of 92.2% and 50.1% was observed in chloroform and methanolic extracts of *Chlorella* Sp. with the LC_{50} of 116.82 and 119.20 ppm.

Larvicidal activity of macroalgae:

Manilal *et al.*, 2011, screened twenty methanolic extracts of marine algae such as *Valoniopsis pachynema, Chaetomorpha antennina, Enteromorpha intestinalis, Acrosiphonia orientalis, Ulva faciata, Caulerpa racemosa, Bryopsis plumose, Dictyota dichotoma, Padina tetrastromatica, Chnoospora bicanaliculata, Sargassum wightii, Stoechospermum marginatum, Lobophora variegata, Spatoglossum asperum, Gracilaria corticata, Hypnea pannosa, Centroceras clavulatum, Cheilosporum spectabile, Portieria hornemannii* and *Gelidium* Sp. They were made into different concentrations such as 100, 200, 300 and 400 $\mu g/L^{-1}$, treated against second and third instar larvae of *Culex quinquefaciatus* and *Aedes aegypti*. Finally, the author concluded that seven macroalgae *Acrosiphonia orientalis, Padina tetrastromatica, Sargassum wightii, Stoechospermum marginatum, Lobophora variegata, Spatoglossum asperum, Centroceras clavulatum* was effective against second instar larvae compared to third instar larvae.

Among the seven species, the highest activity was seen in *Labophora varigata* with the LC_{50} value of 70.38 $\mu g/L^{-1}$ and LC_{90} value of 79.43 $\mu g/L^{-1}$

Khanavi *et al.*, 2011, selected two species of brown algae, *Sargassum swartzii* and *Chondria dasyphylla*, collected from marine water, Iran, extracted using solvents such as n-hexane, chloroform, methyl acetate and methanol. The extractions were subjected to column chromatography and the isolated fractions were made into different concentrations from 2.5 mg/l to 40 mg/l. These fractions were tested against matured third instar and early fourth instar larvae of *Anopheles stephensi* and the maximum activity was found in ethyl extracts of *Sargassum swartzii* and *Chondria dasyphylla* with the LC_{50} value of 11.758 and 10.625; LC_{90} value of 53.472 and 56.394 ppm.

Guedes *et al.*, 2014, investigated larvicidal activity from five different macro algae *Ulva lactuca* MAC51238, *Padina gymnospora* MAC51235, *Sargassum vulgare* MAC51236, *Hypnea musciformis* MAC51234 and *Digenea simplex* MAC51231. They were extracted using different solvents such as dichloromethane, methanol, ethanol and water, tested against fourth instar larvae of *Aedes aegypti*. From the assay, it was concluded that *Padina gymnospora* extracts of chloroform and hexane fractions shown highest activity of 29.018 and 17.230 $\mu g/mL^{-1}$ at LC_{50} and $LC_{90.}$

Bianco *et al.*, 2013, screened 15 species of red, brown and green seaweeds such as *Canistrocarpus cervicornis*, *Dictyota mertensii*, *Dictyopteris delicatula*, *Padina gymnospora*, *Lobophora variegata*, *Sargassum vulgare* var *nanum*, *Sargassum vulgare* var *vulgae*, *Digenia simplex*, *Laurencia dendroidea*, *Palisada perforata*, *Hypnea musciformis*, *Gracilaria* Sp., *Chaetomorpha antennina*, *Dictyosphaeria versluysi*, *Caulerpa racemosa*, extracted dichloromethane and methanol solvents in 2:1 ratio, tested against 20 early fourth instar larvae of *Aedes aegypti*. From the results, 50% mortality was found in *Hypnea musciformis* and *Chaetomorpha antennia* at 300 ppm and 91% mortality found in *Laurencia dendroidea* at 50 ppm and the novel compound elatol was reported.

Thangam and Kathiresan *et al.*, 1991, selected acetone extracts of two seaweeds such as *Caulerpa scalpelliformis* and *Dictyota dichotoma* were mixed with insecticides DDT, BHC and malathion. 12 different concentrations of 0.06, 0.08, 0.10, 0.12, 0.14, 0.16, 0.20, 0.22, 0.24, 0.26 and 0.28 mg L^{-1} were prepared and tested against fourth instar larvae of *Aedes aegypti*, the synergistic activity was found to be 0.89, 0.71 and 0.84 in *Caulerpa scalpelliformis;* 0.97, 0.76 and 0.85 found in *Dictyota dichotoma* at 5 mg L^{-1}.

Bantoto and Danilo Dy, 2013, performed a comparative study on ethanolic extracts of two species of brown marine algae *Padina minor* and *Dictyota linearis*, prepared into different concentrations of 20, 40, 60, 80 and 100 mg/ml, tested against fourth instar larvae of *Aedes aegypti*. From the assay it was found, both the extracts showed 100% mortality at 100 mg/ml, the mortality was moderate at 40 mg/ml and at 20 mg/ml *Dictyota linearis* extracts did not exhibited larvicidal activity but 8% mortality was found in *Padina minor* extracts.

Ali *et al.*, 2013, screened 8 microalgae such as *Ulva lactuca, Caulerpa racemose, Sargassum microystum, Caulerpa scalpelliformis, Gracilaria corticata, Turbinaria decurrens, Turbinaria conoides* and *Caulerpa toxifolia*, extracts were prepared using ethanol and water solvents (3:1), made into different concentrations (10 -100 µg/ml), tested against fourth instar larvae of *Aedes aegypti, Culex quinquefaciatus* and *Anopheles stephensi*. The efficient results were found in *Caulerpa racemosa* against all the three mosquito Species of 0.055 ± 0.010 µg/ml, 0.067 ± 0.136 µg/ml and 0.066 ± 0.007 µg/ml.

Yu *et al.*, in 2015, screened methanolic extracts of three macroalgae such as *Bryopsis pennata, Sargassum binderi* and *Padina australis*. The extractions were further subjected to partition method by combining methanol into hexane, chloroform and aqueous in different concentrations, treated against fourth instar larvae of *Aedes aegypti*. From the above test, it is reported that *Bryopsis pennata* extract with chloroform partition shows a strong larvicidal activity of 82.55 mg/ml at LC_{50}

Vimaladevi *et al.*, 2012, studied the phenolic compounds isolated from *Chaetomorpha antennina* (Bory) Kuetz, collected three phases of phenols such as conjugated soluble phenolic compounds, phenolic free compounds and bound insoluble phenolic acids. About, 2 mg of all the three compounds were mixed with 1ml of methanol, prepared into different concentrations from 10 -100 $\mu g/L^{-1}$, tested against 20 third instar larvae of *Aedes aegypti*, this results shows that that bound insoluble phenolic acid and soluble phenolic acid fraction had an inhibitory effect with a value of 23.4 and 44.6 $\mu g /L^{-1}$ at LC_{50} with a least concentration of 10 $\mu g/ cm^{-2.}$

Raj *et al.*, 2017, screened three macroalgae *Halimeda macroloba, Caulerpa racemosa* and *Ulva lactuca*, extracted using hexane, chloroform, ethyl acetate, acetone and methanol solvents, made into different concentrations of 200, 400, 600, 800 and 1000 ppm, tested against fourth instar larvae. Among the tested species of macroalgae, ethyl extracts of *Caulerpa racemosa* showed maximum larvicidal activity against *Aedes aegypti* with the LC_{50} and LC_{90} value of 579.9 and 1255.4 ppm after 24 hours.

Manilal *et al.*, 2009, collected 13 species such as *Valoniopsis pachynema, Chaetomorpha antennina, Enteromorpha intestinalis, Acrosiphonia orientalis, Ulva fasciata, Caulerpa racemosa, Dictyota dichotoma, Padina tetrastromatica, Chnoospora bicanaliculata, Gracilaria corticata, Hypnea pannosa, Centroceras clavulatum, Cheilosporum spectabile* at Kollam, the extracts were prepared using dichloromethane:methanol (1:1), prepared into different concentrations 100, 200, 300, 400 and 500 $\mu g/ml$, tested against second and third instar larvae. After 24 hours, it was found that *Acrosiphonia orientalis, Padina tetrastomatica* and *Centroceros clavulatum* was most effective against second instar larvae with the LD_{50} value of 91, 96 and 97 $\mu g/ml$ whereas *Acrosiphonia orientalis, Caulerpa racemosa* and *Centroceros Clavatum* was found effective against third instar larvae with the LD_{50} value of 158, 194 and 199 $\mu g/ml$.

Rohani *et al.*, 2016, screened 15 species of methanolic extracts of green macro algae *Dictyota dichotoma, Padina australis, Sargassum binderi, Sargassum polycystum, Sargassum siliquosum, Turbinaria conoides, Bryopsis pennataa, Caulerpa lentilifera, Caulerpa racemosa, Ulva lactuca, Ulva reticulata, Ceratodicyton spongiosum, Gracilaria changii, Gracilaria Salicornia* and *Solieria robusta*, prepared into different concentrations 100, 200, 300, 400 and 500 $\mu g/ml$, tested against 25 third instar larvae of *Aedes aegypti* and *Aedes albopictus*.

From the results, *Bryopsis pennata* showed maximum larvicidal activity against *Aedes aegypti* and *Aedes albopictus* with the LC_{50} value of 156.97 µg/ml and 177.50 µg/ml.

Antonysamy *et al.*, 2015, screened petroleum ether, chloroform, acetone, ethyl acetate, ethanol and distilled water solvents extracts of a brown algae *Dictyota bartayresiana*, made into different concentrations from 50, 100, 150, 200 and 250 mg/L tested against fourth instar larvae of *Culex quinquefasciatus*. It showed mortality with the LC_{50} and LC_{90} value of 166.33 and 265.69 mg/L.

Nagaraj and Osborne in 2014, investigated methanolic extracts of green algae *Caulerpa racemosa*, prepared into different concentrations ranging 80 - 400 mg/L, tested against 20 early fourth instar larvae of *Culex trianiorhynchus*. After 24 hours, all the larvae were died and the compounds responsible for lethality was identified as 2-(-3-bromo-1-adamantyl) acetic acid methyl ester and chola-5, 22-dien-3-ol through GC-MS.

Poonguzhali and Nisha, 2012, screened methanol, acetone and benzene extracts of two seaweeds, *Ulva fasciata* and *Grateloupia lithophila*, prepared into different concentrations 100, 200, 300, 400 and 500 mg/L tested against *Culex quinquefasciatus*. It was found that all the three extracts of *Grateloupia lithophila* was found to be effective with the LC_{50} value of 431.90, 349.74 and 425.42 ppm respectively.

Orlando *et al.*, 2016, isolated red algae *Laurencia dendroidea*, collected from two different places "Azeda" beach and vermelha beach, extracts were prepared into different concentrations ranging from 1.25 ppm to 10 ppm. The compounds which shown maximum larvicidal activity further fractionated and undergone GC-MS analysis. The results showed that halogenated compound (-) elatol was found in the Species collected at "Azeda" beach whereas the same compound with one more significant compound (+)- obtusol was found in the same species collected from "vermelha" beach.

Abou-Elnaga *et al.*, 2011, investigated the petroleum ether extracts of *Laurencia papillosa*, undergone fractionation and characterization. Four compounds, C_{15} acetogenin, (12Z)-transmaneonene-B, sesquiterpenoid 2,10-dibromo-3-chloro-a-chamigrene and fatty acid aldehydes was isolated, treated against first, second, third and fourth instar larvae of *Culex pipiens*. After 24 hours, novel compound acetogenin showed highest lethal rate with the LC_{50} value of 30.7, 36.9 and 41.8 ppm against second, third and fourth instar, respectively.

Kumari *et al*., 2020, studied the ethanol, acetone, benzene and diethyl ether extracts of brown algae *Turbinaria conoides* collected from Mandapam coast, Tamil Nadu, tested against *Culex quinquefasciatus*, the result from the bioassay showed potential activity in diethyl ether extract with the LC_{50} at 104 mg/L, acetone extracts at 469 mg/L, benzene extract at 781 mg/L and ethanol extracts at 1917 mg/L, respectively.

Balaraman *et al*., 2020, developed a stable silver nanoparticles through aqueous extracts of *Sargassum myriocystum*, confirmed through UV-Vis spectroscopy, X- Ray diffraction, FTIR, TEM and SEM was prepared in to different concentrations from 0 - 25 mg/L, tested against the fourth instar larvae of *Aedes aegypti* and *Culex quinquefasciatus* and the results from the analysis showed *Sargassum myriocystum AgNPs* found efficient against *Culex quinquefasciatus* with the LC_{50} value of 19.31 mg/L and 5.59 mg/L after 24 hours and 48 hours, respectively.

Deepak *et al*., 2016, synthesized silver nanoparticles from *Turbinaria arnata* (To-*AgNPs*) characterised by UV-Vis spectroscopy, X ray diffraction, FTIR, EDX, FE-SEM, was prepared in to different concentrations from 0.0625-10 ppm, tested against fourth instar larvae of *Aedes aegypti, Anopheles stephensi* and *Culex quinquefasciatus*. The result from the bioassay revealed the highest mortality was found in *Aedes aegypti* > *Anopheles stephensi* > *Culex quinquefasciatus* with the LC_{50} value of 0.738, 1.134 and 1.494 µg/ml; LC_{90} value of 3.342, 17.982 and 22.475 µg/ml.

Kalimuthu *et al*., 2013, studied the effect of acetone, methanolic and petroleum ether extracts of *Gracillaria firma* combined with *megacyclops formosanus* was tested against first to fourth instar larvae of *Aedes aegypti,* the result from the bioassay reveals the highest mortality was recorded in methanolic extracts against *Aedes aegypti* with the LC_{50} value of 0.251%

Deepak *et al.*, 2019, investigated the crude methanolic extracts of *Halymenia palmate* followed by the fractions were collected using column chromatography with the different ratios of petroleum ether and ethyl acetate. The fractions (Hpf-1 and Hpf-2) collected, was prepared into different concentrations from 50-300 ppm tested against first, second and third instar larvae of *Aedes aegypti,* the result from the analysis showed the higher activity was found in Hpf-2 with the LC_{50} and LC_{90} of 23.69 and 233.49 µg/ml.

Larvicidal activity of Plants:

Al-Solami, 2021, screened the acetone extracts of four different plants *Lantana camara, Rhazya stricta, Acalypha fruticose* and *Ruta chalepensis* against *Culex pipiens*. The results showed 98% mortality against *Lantana camara,* 91% mortality was observed in *Rhazya stricta,* 79% mortality was seen in *Acalypha fruiticosa* and 69% mortality was observed in *Ruta chalepensis*. Since, *Lantana camera* was found efficient, their level of *Ache* enzyme was further studied in qt RT- PCR and was found high.

Filho *et al.*, 2021, screened essential oils of ten piper Sp. *Piper aduncum, Piper marginatum, Piper gaudichaudianum, Piper crassinervium, Piper arboreum, Piper hemmendorffii, Piper cernuum, Piper lucaenum, Piper lindbergii* and *Piper amalago* tested against *Aedes aegypti,* the results from the analysis showed six species of *Piper aduncum, Piper marginatum, Piper gaudichaudianum, Piper crassinervium* and *Piper arboreum* found up to 100% mortality at 100 ppm. The active compounds responsible for the larvicidal activity was found as phenyl propanoids such as b-Asarone, (E)-Anethole, (E)-b-Caryophyllene, monoterpenes such as g-Terpinene, p-Cymene, Limonene, a-Pinene, and b-Pinene.

Akhal *et al.*, 2021, examined the essential oils of *Lavandula angustifolia* and *Lavandula dentate* Collected from Morocco, tested against *Culex pipens* and the results from the analysis reveals *Lavandula angustifolia* was found efficient with the LC_{50} and LC_{90} value of 140 - 450 µg/ml followed by *Lavandula dentate* with the LC_{50} and LC_{90} value of 2670 µg/ml and 7400 µg/ml, the chemical analysis reveals the presence of linalool, linalyl acetate, geraniol, lavandulyl acetate, camphor, β-caryophyllene, terpinen-4-ol, β-myrcene in *lavandula angustifolia* and 1,8-cineole, camphor, α-pinene, transpinocarveol, linalool and borneol found in *Lavandula dentate.*

Carneiro *et al.*, 2021, extracted with 90% methanolic extracts of *Eugenia astringens*, *Myrrhinium atropurpureum* and *Neomitranthes obscura* leaves, collected from Marambaia and Grumari restingas, tested against *Aedes aegypti* and *Simulium pertinax*, the results reveal that *Myrrhinium atropurpureum* has an efficient larvicidal activity of 50% and 100% against *Aedes aegypti* with the LC_{50} value of 11.10 and 9.68 ppm after 24 hours and 48 hours, respectively whereas, *Eugenia astringens* showed mortality of 50% and 63.33% after 24 hours and 48 hours exposure, *Neomitranthes obscura* exhibited mortality of 46.66% after 48 hours against *Aedes aegypti* and *Simulium pertinax* larvae showed 100% mortality after 24 hours exposure.

Raghavendra *et al.*, 2021, extracted the leaves of two medicinal plants *Rubus steudneri* and *Rubus apetalus* using methanol through maceration process, they were treated against the second and third instar larvae of *Aedes* and *Culex* Sp. Among the study, it was observed that second and third instar larvae of *Culex* Sp. was susceptible towards both the leaf extracts with the LC_{50} and LC_{90} value of 0.144 and 0.296; 0.157 and 0.378 mg/ml, respectively after 24 hours.

Gunathilaka *et al.*, 2021, analysed the four different morphologies of zinc oxide nano particles of star (S), needle (N), plate (P) and cubical (C) shaped was prepared into different concentrations from 25 -100 mg/L, tested against *Aedes albopictus* and *Anopheles vagus*. From the analysis it was found that the highest mortality was observed in the Zinc *NPs* of Star shaped against *Aedes aegypti* and *Anopheles vagus* with the LC_{50} value of 38.90 and 4.78 mg L, the least mortality was observed in Zinc *NPs* of plate like against *Aedes aegypti* and *Anopheles vagus* with the LC_{50} value of 68.38 and 13.64 mg L.

Hadi *et al.*, 2021, macerated the leaves of *Evodia suaveolens* using n-hexane solvent, tested against third instar larvae of *Aedes aegypti* with 1 ppm concentrations with ten replications, the percentage of mortality was observed as 11.6%.

Silva *et al.*, 2021, extracted ethanolic extracts of *Azadirachta indica*, prepared into different concentration from 250 - 2000 µg/mL treated against third instar larvae of *Aedes aegypti*, the result from the analysis showed higher mortality with the LC_{50} of 10% higher than the control ranging from 12-46%.

Betim *et al*., 2021, examined *Ocotea nutans* leaves and stems extracted using hydoethanolic solvent and it was fractionated with hexane, chloroform and ethyl acetate in the increasing polarity of solvents and tested against third instar larvae of *Aedes aegypti*, from the bioassay it reveals that hexane fractions of stem found more efficient with the LC$_{50}$ of 14.14 µg mL^{-1} after 24 hours.

Gomes *et al*., 2021, screened four different types of *Aedes aegypti* resistant to deltamethrin and temephos was tested with Yeast encapsulated orange oil (YEOO) from *Citrus sinensis* and some other assays were performed with Belo horizonte strain using different environmental factors such as natural light and temperature. YEOO exhibited LC$_{50}$ of 8.1 – 24.7 mg/L against *Aedes aegypti* strains tested in laboratory conditions and it was suggested that YEOO can be a good alternative for synthetic insecticides.

Torawane *et al*., 2021, screened the ethanolic leaf extracts of four plants *Cyathocline purpurea, Blumea lacera, Neanotis montholonii* and *Neanotis lancifolia,* prepared into four different concentrations 0.05, 0.1, 0.2 and 0.3 mg/ml, tested against third instar larvae of *Aedes aegypti,* the highest mortality was observed in *Cyathocline purpurea* with the LC$_{50}$ of 0.088 mg/ml followed by *Blumea lacera* extract with the LC$_{50}$ of 0.758 mg/ml after 72 hours, respectively.

Nakasen *et al*., 2021, screened the efficacy of compound cinnamaldehyde from *Cinnamum verum* essential oil prepared into different concentrations from 12.5, 25 and 50 ppm tested against *Culex quinquefasciatus*. The larvicidal activity of cinnamaldehyde was effective in killing 100% larvae at 25 and 50 ppm, after 72 hours. It was suggested that it is green friendly and good alternate for synthetic insecticides in endemic areas.

Dinesh *et al*., 2021, extracted the aqueous, acetone and methanolic extracts of *Salvia leucantha* plant leaves made into different doses from 100 µg/mL to 500 µg/mL, tested against fourth instar larvae of three mosquito vector *Anopheles stephensi, Aedes aegypti* and *Culex quinquefasciatus.* The enzyme levels of total proteins, acetylcholinesterase, α- and β- carboxylesterase was performed, the results showed that 100% mortality was found in all the extracts and it was suggested as a potent toxic agent that strongly reduces mosquito larvae.

Ravindran *et al*., 2020, studied the aqueous and methanolic extracts of *Clitorea ternatea* flower, tested against early fourth instar larvae of *Aedes aegypti* and *Aedes albopictus*. From the larvicidal bioassay, it was found that the highest mortality was found against *Aedes aegypti* with the LC_{50} and LC_{95} values of 1056 and 2491 mg/L, respectively and against *Aedes albopictus* with the LC_{50} and LC_{95} values of 1425 and 2753 mg/L.

Dinh *et al*., 2020, studied the infusion of hot water and ethanolic extracts of *Sollanum nigrum*, prepared in to different concentrations from 1.25% - 20%, tested against the laboratory strains of *Aedes aegypti* and *Aedes albopictus*, the results showed that the hot water infusions of *Sollanum nigrum* showed the mortality up to 77.3% after 72 hours post exposure and 100% mortality was observed in ethanolic extracts of *Sollanum nigrum* against *Aedes aegypti* and *Aedes albopictus*.

Posse *et al*., 2020, studied the essential oil of *Hypenia irregularis* extracted by hydro distillation, tested against third instar larvae of *Aedes aegypti*. The chemical composition of the GCMS shows 2,5-dimethoxy-ρ-cymene, thymol, *o*-cymene, phenol-3-(1,1-dimethylethyl)-4-methoxy and humulene with 27.0, 21.36, 15.56, 8.89 and 5.01%, respectively and it was reported that the action of essential oil of *Hypenia irregularis* showed better results than the commercial insecticides.

Manzano *et al*., 2020, evaluated the ethanolic extracts of *Azadirachta indica* leaves made into different concentrations 10 mg L^{-1}, 20 mg L^{-1}, and 50 mg L^{-1}. From the assay, the higher larval mortality of 93% was observed at a concentration of 50 mg L^{-1}; 47% and 70% was observed at 10 mg L^{-1} and 20 mg L^{-1}, respectively after 72 hours of exposure and the major compound detected by the GC-MS analysis was phytol with the area percentage of 14.4.

Murugammal *et al*., 2020, screened the *n*-hexane, chloroform, ethyl acetate and ethanol solvent extracts of *Epaltes pygmaea* was prepared into different concentrations 62.5 ppm, 125 ppm, 250 ppm and 500 ppm tested against third instar larvae of three mosquito vector *Culex quinquefasciatus, Aedes aegypti* and *Anopheles stephensi*. The result from the bioassay reveals that ethyl acetate extract was found most effective with the LC_{50} values of 29.94 ppm,

35.79 ppm and 62.37 ppm against *Anopheles stephensi*, *Culex quinquefasciatus* and *Aedes aegypti*, respectively.

Ersino *et al.*, 2020, evaluated the chloroform, petroleum ether, ethanol and acetone extracts of *Juniperus procera,* tested against twenty third and fourth instar larvae of *Anopheles arabiensis,* the result showed that the mortality rate increased as the concentration increases in the following order petroleum ether < chloroform < acetone < ethanol extracts with respect to its mean ± SD values.

Meisyara, 2020, studied the crude methanolic extracts of stem bark, leaves, seeds, kernels and rinds of *Cerbera odollam* and the fractionation was done using n-hexane, ethyl acetate and distilled water using different concentrations from 250 - 2000 ppm followed by the sub–fractionations carried out using ethyl acetate at 1000 ppm, tested against second instar larvae of *Culex quinquefasciatus,* was found ethyl acetate and n-hexane showed highest larvicidal activities of seed kernel after 24 and 48 hours of treatment.

Ombugadu *et al.*, 2020, studied the larvicidal activity of *Capsicum chinensis* collected from farm lands, Nigeria, tested against larvae of *Culex quinquefasciatus* and *Anopheles gambiae*, it was observed to be susceptible against *Anopheles gambiae* with the LD_{50} and LD_{90} values of 21.88 mg/mL and 52.48 mg/mL at 24 hours respectively; 16.98 mg/mL and 38.90 mg/mL at 48 hours respectively; 13.80 mg/mL and 30.19 mg/mL at 72 hours, respectively while *Culex quinquefasciatus* doesn't show activity.

Amarasinghe *et al.*, 2020, synthesized green silver nanoparticles from *Annona glabra* characterised by UV-Vis spectroscopy, SEM, TEM, Dynamic Light scattering technique, FTIR and phytochemical test were all confirmed the presence of $AgNO_3$. The crude extracts and An-$AgNO_3$ were tested against the third instar larvae of *Aedes aegypti* and *Aedes albopictus* using the concentrations from 2–10 mg/L. Results from the bioassay reveals that the An-$AgNO_3$ was found to be superior and more susceptible to the *Aedes albopictus* than the *Aedes aegypti* with the LC_{50} value of 3.02 mg/L and 2.51 mg/L.

Shehata *et al.*, 2020, studied the larvicidal activity of ethanolic, petroleum ether and chloroform extracts of *Pulicaria jaubertii* leaves, made into different concentrations between 25 to 150 ppm, tested against laboratory reared *Aedes aegypti*, from the bioassay it was found 100% larval mortality at 50 and 75 ppm only in ethanolic extracts.

Souheila *et al.*, 2020, examined the larvicidal activity of ethanolic and aqueous extracts of *Tamarix aphylla* (Family: Tamaricaceae), *Jacaranda mimosifolia* (Family: Bignoniaceae), *Lavandula angustifolia* (Family: Lamiaceae), prepared into different concentrations from 1-20 mg mL^{-1} tested against the late second, third and fourth instar larvae of *Culex quinquefasciatus*, the result showed the ethanolic extracts of *Tamarix aphylla* found highest mortality of 100% after 72 hours. The phytochemical analysis showed the presence of flavonoids, alkaloids, tannins and polyphenols.

Famuyiwa *et al.*, 2020, screened the methanolic extracts of fifteen plants, *Anthocleista vogelli* (Gentianaceae), *Calotropis procera* (Asclepiadaceae), *Cassia sieberiana* (Fabaceae), *Delonix regia* (Fabaceae), *Ficus exasperata* (Moraceae), *Ficus sur* (Moraceae), *Ficus vogelli* (Moraceae), *Hilleria latifolia* (Petiveriaceae), *Momordica charantia* (Cucurbitaceae), *Psidium guajava* (Asclepiadaceae), *Senna alata* (Fabaceae), *Solanum macrocarpon* (Solanaceae), *Spondias mombin* (Anacardiaceae), *Thevetia nerifolia* (Apocynaceae), *Vitex doniana* (Verbanaceae) tested against fourth instar larvae of *Culex quinquefasciatus*, n-hexane and ethylacetate partitioned fraction was also tested against the fourth instar larvae of *Culex quinquefasciatus*. The results from the bioassay showed that the leaf extracts of *Calotropis procera* and *Solanum macrocarpon* and fruit extracts of *Thevetia neriifolia* was active against *Culex quinquefasciatus*; the non-polar solvent, n-hexane fractions of *Solanum macrocarpon* and *Spondias mombin* was most active with the LC$_{50}$ value of 0.78 ± 0.03 mg/mL and 0.81 ± 0.03 mg/mL against *Culex quinquefasciatus*.

Chaves *et al.*, 2020, studied the larvicidal activity of essential oil of *Origanum majorana* evaluated against the third instar larvae of *Aedes aegypti*, the results from the analysis showed higher mortality against *Aedes aegypti* and the chromatographic studies show the major compound as pulegone (57.05%), verbenone (16.92%), trans-p-menthan-2-one (8.57%), iso-menthone (5.58%), piperitone (2.83%), 3-octanol (2.35%) and isopulegol (1.47%).

Zulhussnain *et al.*, 2020, screened the five different weed plants such as *Convulvulus arvensis, Chenopodium murale, Tribulus terrestris, Trianthema portulacastrum* and *Achyranthes aspera* tested against *Culex quinquefasciatus,* from the analysis it was reported that *Achyranthes aspera* showed 100% mortality at 250 ppm with the LC_{50} of 87.46, 39.08 and 9.22 ppm at 24, 48 and 72 hours, respectively. The Phytochemical analysis showed the presence of flavonoids, saponins, tannins, steroids, cardiac glycosides, alkaloids, anthrequinones and terpenoids; FTIR analysis showed the presence of phenolic compounds and was suggested as the reason for death of larvae of *Culex quinquefasciatus.*

Nhaca *et al.*, 2020, studied the crude aqueous, ethanolic extracts and the essential oils of *Annona muricata, Annona squamosa, Annona senegalensis, Lantana camara, Strychnos henningsii, Strychnos madagascariensis, Strychnos spinosa, Sclerocarya birrea, Securidaca longepedunculata, Eucalyptus citriodora* and *Cymbopogon citratus,* collected from Southern Mozambique, evaluated against third instar larvae of *Anopheles arabiensis,* after the 24 hours of exposure it was found that ethanolic extracts showed higher efficacy against *Anopheles arabiensis.* The Phytochemical screening revealed the presence of flavonoids, steroids, triterpenoids, saponins and tannins in ethanolic extracts.

Umadevi and Jaleel *et al*, 2020, studied the larvicidal activity of essential oil of leaves of *Gliricidia sepium,* prepared into different concentrations from 10 - 250 mg/ml, tested against the fourth instar larvae of *Aedes aegypti.* The results show that the aqueous extract of *Gliricidia sepium* showed the potential larvicidal activity of essential oil at 250 mg/ml.

Nityasree *et al.*, 2020, evaluated the petroleum ether, methanol and aqueous extracts of *Solanum lycopersicum* leaf, made in to different concentrations from 10 - 50 mg/ml, tested against third instar larvae of *Aedes aegypti.* The results showed methanolic extracts has a strong larvicidal activity after 48 hours of treatment with the LC_{50} value of 20.323 mg/ml, the phytochemical analysis revealed the presence of alkaloids, saponins, phenols and flavonoids. The major compounds observed through GCMS was phytol acetate (42.66%), neophytadiene (29.38%), the growth disruption assay showed abnormal coiling and movement in the larvae after 24 hours and histopathological studies revealed a damage in the digestive and respiratory tract of *Aedes aegypti* larvae after 48 hours.

Al-Azab *et al.*, 2020, investigated the acetone extracts of four different medicinal plants *Rhazya stricta, Lantana camara, Ruta chalepensis,* and *Punica granatum* leaves and peels, prepared into different concentrations from 150 - 1000 ppm, tested against late third and early fourth instar larvae of *Aedes aegypti*. The result from the bio assay reveals the mortality ranges from 25 - 97% against *Rhazya stricta*, 23 - 94% mortality against *Lantana camara*, 17 - 96% against *Ruta chalepensis* and 10 - 72% against *Punica granatum*.

Timothy, 2020, studied the methanolic extracts of *Azadirachta Indica,* prepared into different concentrations from 0.4 -1.6 ml, tested against third and fourth instar larvae, checked at the regular intervals of 6, 12 and 24 hours, the results showed the *Azadirachta indica* was toxic even at low dosages with the LC_{50} value of 3.02 mg/ml; LC_{90} value of 33.11 mg/ml.

Muniandy *et al.*, 2020, investigated the larvicidal activity of decoction of *Citrus aurantifolia* against the larvae of *Aedes aegypti* and the highest mortality was observed at 30% concentration with the LC_{50} value of 38.5% and 6.6% after 24 and 48 hours, respectively. The LC_{60} value of 91.6% and 64.4% after 24 and 48 hours, respectively, However, LC_{90} could not be determined.

Ravindran *et al.*, 2020, investigated the aqueous, methanol and ethyl acetate flower extracts of *Clitoria ternatea* tested against fourth instar larvae of *Aedes aegypti and Aedes albopictus.* The maximum larvicidal activity was observed against *Aedes aegypti* with the LC_{50} and LC_{95} values of 1056 and 2491 mg/L, respectively and against *Aedes albopictus* with the LC_{50} and LC_{95} values of 1425 and 2753 mg/L. The non-target effect of *Poecilia reticulata* possess non-toxic upto 2500 mg/L. The major compounds found in GCMS analysis was Glycerin, 2-Hydroxy-gamma-butyrolactone, Neophytadiene, n-Hexadecanoic acid, cis-Vaccenic acid and Octadecanoic acid with the total percentage of 28.7 and the phytochemical analysis showed presence of phenols such as anthocyanin, flavonoid and tannin.

Chaiphongpachara *et al.*, 2020, screened the essential oils of seven herbs such as cassia, cinnamon, East Indian lemongrass, bay, sweet basil, holy basil and ginger against the larvae of *Aedes aegypti*. The results from the bioassay reveals the highest larvicidal activity was found in cinnamon oil with the LC_{50} value of 0.03 ppm; LC_{90} value of 0.04 ppm.

Laojun and chaiphongpachara, 2020, compared the efficacy of essential oils of *Rosmarinus officinalis*, *Vanilla planifolia* and *Mentha spicata* tested against the larvae of *Aedes aegypti*. The results showed efficient larvicidal activity of *Rosmarinus officinalis*, *Vanilla planifolia* and *Mentha spicata* with the LC_{50} of 0.23, 0.10, and 0.12 ppm after 24 and 48 hours, respectively.

Ayinde *et al.*, 2020, extracted the oil seed kernel of *Azadirachta indica* collected from Ibadan, Nigeria using the n-hexane, was prepared in to different concentrations from 100 to 500 ppm, tested against the fourth instar larvae of *Anopheles gambiae,* the results from the larvicidal activity found as 91.6 - 100.0% with the LC_{50} and LC_{90} value of 1666.86 and 2880.94 ppm, respectively after three days of exposure at 500 ppm.

Shehu *et al.*, 2020, synthesized the copper-cobalt bimetal nanoparticles from Palmyra palm fruit extracts, characterised by UV-Vis spectrophotometry, FTIR spectroscopy was tested against the first, second and third instar larvae of *Culex quinquefasciatus*. The lethal concentration values of LC_{50} and LC_{90} was found as 12.036 and 143.316; LC_{50} and LC_{90} values of 14.774 and 263.456 against second instar larvae; LC_{50} and LC_{90} value of 16.076 and 296.758 ppm against third instar larvae, respectively.

Udhayashri *et al.*, 2020, studied the chloroform and methanolic extracts of *Gloriosa superba* (Family: Liliaceae), prepared in to different concentrations from 62.5 – 1000 ppm, tested against the mosquito vectors of *Culex quinquefasciatus, Anopheles stephensi, Aedes aegypti* and *Aedes albopictus.* The results from the bioassay showed that the methanolic extract showed 100% mortality against *Culex quinquefasciatus* and *Anopheles stephensi,* 98% mortality against *Aedes aegypti* and 96% mortality against *Aedes albopictus* at 1000 ppm concentration, respectively.

The phytochemical analysis of methanolic extracts showed the presence of alkaloids, saponins, tannins, flavonoids, phenols and steroids.

Wahedi *et al.*, 2020, evaluated the ethanol and aqueous extracts of *Azadirachta indica*, prepared in to different concentrations from 50, 100 and 200 mg/ml, tested against the second and third instar larvae of *Anopheles* Sp. the results from the bioassay reveals that ethanolic extracts of *Azadirachta indica* were found more efficient with the 100% mortality. The Lethal concentrations of LC_{50} and LC_{95} value was found as 12.309 and 3.589.

Viswan and Pushapalatha, 2020 studied the acetone and methanolic extracts of *Anamirta cocculus* seeds collected from the Calicut university campus was tested against *Culex quinquefaciatus* and *Aedes albopictus*. The methanolic extract was also defatted with equal volume of petroleum ether and fractionised using ethyl acetate and double distilled water. The result from the bioassay of *Anamirta cocculus* showed higher activity against the *Aedes albopictus* than the methanolic extracts.

Kazempour *et al.*, 2020 investigated the essential oils of native plants such as *Artemisia sieberi* and *Tanacetum balsamita*, collected from different localities of Iran, tested against *Anopheles stephensi*. The results from the bioassay reveals that *Tanacetum balsamita* showed higher larvicidal activity with the LC_{50} and LC_{90} value of 26.2 and 52.4 ppm. The GC-MS analysis showed the presence of 39 constituents and the major compound was found as Thujone (52.37%) and Carvone (26.84%).

Ngwu *et al.*, 2020, evaluated the methanolic extracts of *Ipomoea stolonifera*, prepared into different concentrations from 50 mg/L – 5000 mg/L, tested against the fourth instar larvae of laboratorically reared *Aedes aegypti* and the result from the analysis found with the LC_{50} of 1.24 ± 0.09 and it was further recommended to find the active compound in the efficient control of mosquito larvae.

Govindan *et al.*, 2020, synthesized the silver nanoparticles (*AgNPs*) from the aqueous extracts of *Plumbago auriculata* characterised through FTIR, XRD, TEM, EXD, Zeta Potential and DLS, evaluated against the fourth instar larvae of *Aedes aegypti and Culex quinquefasciatus*. The result from the bioassay proved to inhibit the fourth instar larvae of *Aedes aegypti* and *Culex quinquefasciatus* at the concentrations of 45.1 mg/mL and 41.1 mg/mL.

Waris *et al.*, 2020, studied the crude *AgNPs* leaf extracts of *Ricinus communis*, characterised through UV-Vis spectroscopy, FTIR, XRD, prepared into different concentrations from 50-250 ppm made into five replicates, tested against second and third instar larvae of *Aedes aegypti*. The result from the analysis show the *AgNPs* synthesised nanoparticles found more toxic with the LC_{50} and LC_{90} values of 46.22 ppm and 85.30 ppm than the plant extracts.

Sandhya *et al.*, 2021, studied the larvicidal bioassay of *Vitex negundo*, tested against early third instar larvae of *Aedes aegypti, Aedes albopictus* and *Culex quinquefasciatus*, the results showed highest mortality of 100% found against third instar larvae of *Culex quinquefasciatus* than *Aedes* Sp. The EC_{50} values against *Culex quinquifasciatus, Aedes aegypti* and *Aedes albopictus* was found as 83.77 ppm, 341.7 ppm and 487.9 ppm, respectively.

Daniel *et al.*, 2020, isolated two novel phenolic compound such as vanillic acid and protocatechuric acid from *Koschya thymodora* are active against *Anopheles gambiae* and *Culex quinquefasciatus*. but, the highest activity was found in *Anopheles gambiae* with the LC_{50} of 77.35 and 62.4 µg/ml after 48 hours exposure.

Tabanca *et al.*, 2018, isolated betasitosterol from *Veratrum lobelium* exhibited the highest activity against *Aedes aegypti* with the LC_{50} of 1.7 ppm. However, in the present study stigmasterol, campesterol and betasitosterol (1:1:1) isolated from *Cymodoceae serrulata* were found effective against third instar larvae of *Culex quinquefasciatus* and *Aedes aegypti* with the LC_{50} value of 0.004 µg/ml and 0.040 µg/ml. They can be comparable with temephos with the LC_{50} value of 0.001 and 0.020 µg/ml, respectively.

acel receptor of mosquito larvae as drug target:

Acetylcholine is a neurotransmitter and acetylcholinesterase (*AChE*) is a serine hydrolase enzyme that belongs to protein family cholinesterase (Sussman *et al.*, 1991). It encodes two genes *ace1* and *ace2* in all insects by its gene duplication (Weill *et al.*, 2002). It was also reported that due to the evolution one of the duplicated gene has lost in some species (Kim and Lee, 2013; Hoffmann *et al.*, 1992), so there is only one '*ace1*' gene present in humans and *Drosophila melanogaster* (Fournier *et al.*, 1989). In the case of mosquito, two genes have been expressed but one gene '*ace1*' appear to have an amino acid sequence identity and central catalytic functions, another one lost its function (Kim and Lee, 2013). During neurotransmission, *AChE* enzyme released into synaptic cleft and binds to *AChE* receptor (Post-synaptic membrane) passes the signals to the downstream nerve cells and neuromuscular junctions to hydrolase acetylcholine. Since, it is inhibited by external source, this lead to paralysis and death of an organism (Berg *et al.*, 2012). Moreover, *AChE* enzyme is the target protein for many drug development, natural biotoxins and some insecticides particularly organophosphates and carbamates (Pohanka., 2011).

Bioinformatics:

Govindan *et al.*, 2020, implemented the molecular docking to determine the binding efficacy of ligand plumbagin by retrieving in the Pubchem database and the D7 Salivary Protein of *Aedes aegypti* retrieved from Protein Data Bank ID No:3DXL and OBP (Odorant Binding Protein) of *Culex quinquefasciatus* from Protein Data Bank ID No:3OGN was retrieved from active site of the drug target. The molecular docking was performed using Auto Dock 4.2 program compiled and run under Microsoft Windows XP operating system. The result shows that plumbagin exhibited binding affinity towards D7 Salivary Protein and Odorant Binding Protein.

Sandhya *et al.*, 2021, analysed the *in silico* docking interaction between the phyto compound Vitexicarpin from *Vitex negundo* against Odorant binding protein of *Culex quinquifasciatus* retrieved from Protein Data Bank ID No: 2L2C using Molegro Virtual Docker 6.0 which has already showed the best results in *in vitro* studies. The results of moldock scoring of Vitexicarpin was found as -87.3733 and the control Azadirachtin was found to be -110.77.

From the above literature, the reader can be able to understand the various researches of phytochemical screening, spectroscopic studies, docking studies, larvicidal activity of microalgae, macro algae, angiosperms and few preliminary works of seagrasses, with the progressing upgrades in the scientific finding on seagrasses against deadly vectors, *C. quinquefasciatus* and *Ae. aegypti* will provide a new approach to the researchers to find an alternative chemical insecticide in future.

4. MATERIALS AND METHODS

4.1. Sample collection, extraction and preliminary phytochemical analysis of seagrasses:

4.1.1. Sample collection, processing and authentication:

Five species of seagrasses *Enhalus acoroides* (L.f) Royle (Fig.2), *Halophila ovalis* (R.Br) Hook.f (Fig.3), *Syringodium isoetifolium* (Asch.) Dandy (Fig.4), *Halodule uninervis* (Forssk.) Boiss (Fig.5) and *Cymodocea serrulata* (R.Br) Asch & Magnus (Fig.6) were selected and collected from the deep bottom of the sea tethered to the mud in the Mandabam Coast, Ramanathapuram District (Fig.7a; Fig.7b). The bulk samples were washed three times with tap water to eliminate all dirt, sand particles and microorganisms. They were chopped into small pieces and allowed to dry under the shade for 7-10 days (Fig.7c). The collected species of seagrasses were pressed in herbarium sheets and authenticated by Dr. G. Gnanasekaran, Taxonomist, Madras Christian College, Tamil Nadu, India. The herbarium sheets are maintained in Pachaiyappa's College with the voucher numbers SGEAPC01, SGHOPC02, SGSIPC03, SGSHUC04 and SGCSPC05 respectively.

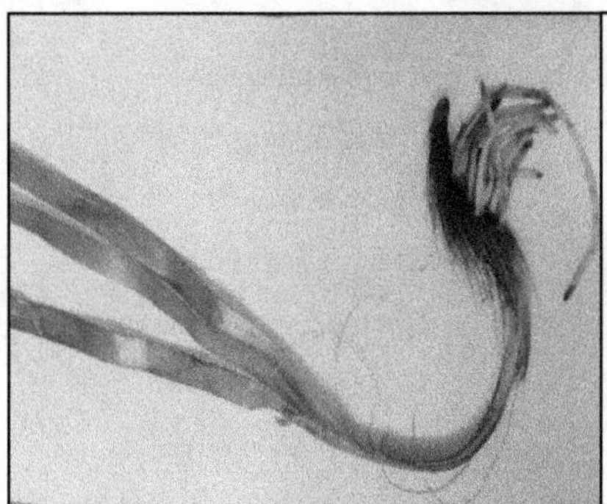

Fig. 2 *Enhalus acoroides* (L.f) Royle

Enhalus acoroides is commonly referred as 'tape seagrass', a dioecious plant with 25 to 300 shoots and a lateral rhizome. It reproduces asexually by rhizome and sexually through male inflorescence, producing pollen, releasing into the water surface and reaching the female inflorescence (Rollen,1998). Pollination occurs during march to august (Brouns and Heijs, 1986) and the flowering period starts between September to November (Troll,1931).

Fig. 3 *Halophila ovalis* (R.Br) Hook.f

Halophila ovalis is commonly referred as 'spoon seagrass' or 'paddle weed', develop in shallow intertidal regions. It has oval shaped or oblong leaf shape arranged over smooth rhizomes. It has separate female and male plants. Female flowers are produced at the base of the shoot, extend up to top to reach male flowers. Male flowers do not bloom often and are less in number. The fruits are colourless and tiny (Waycott *et al.*, 2004; Calumpong and Menez, 1997; Huisman, 2000)

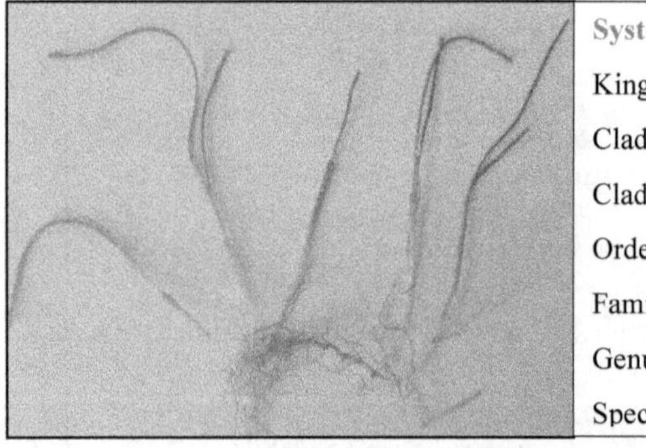

Fig 4 *Syringodium isoetifolium* (Asch.) Dandy

Syringodium isoetifolium is a noodle seagrass found in intertidal regions with increased nutrient, they have tubular leaves with pointed tip, rhizome appears to be slender and has separate male and female flowers, because of the presence of air cavities in the leaves, they can detach from the plant and float in water as it gets older and brittle. Flowering is branched (Cymose inflorescence) and seed set begins from january to late march, second set period begins in November (McKenzie *et al.*, 2016; Waycott *et al.*, 2004; Calumpong and Menez,1997; Hsuan Keng *et al.*, 1998).

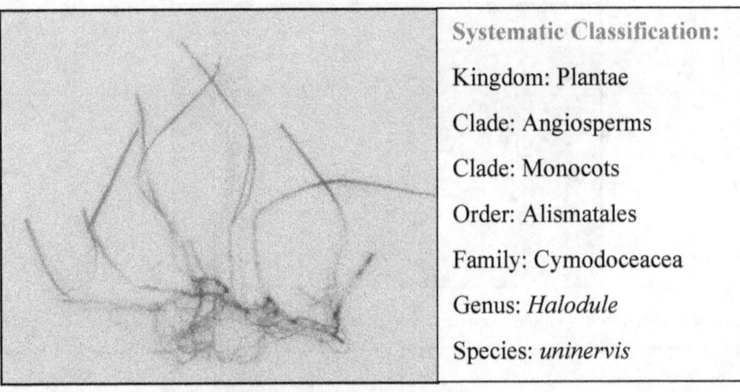

Fig 5 *Halodule uninervis* (Forssk.) Boiss

Halodule uninervis is commonly called as 'needle seagrass', narrow, flat like a ribbon shaped leaves, dioecious plants with wiry and erect stem, flowering is perennial. They vegetatively reproduce by diverse spread of leaves and sexually by male flowers which grows at the base of leaf sheath for only short span of time. Flowering and fruiting takes place between June-July. Seeds are very hard and released into the water currents (West *et al.*, 1998; Jupp *et al.*, 1996; Den Hertog, 1964)

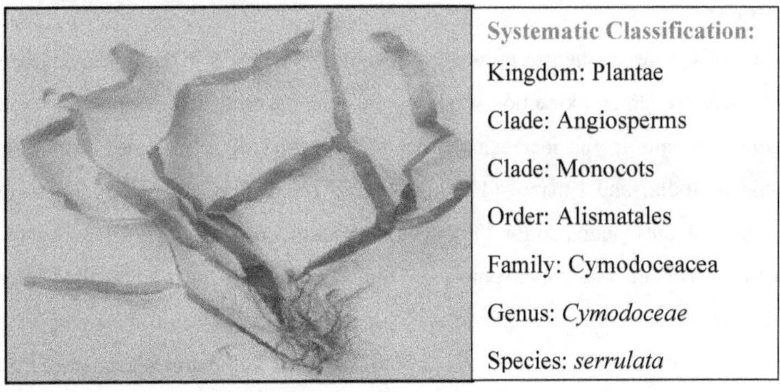

Fig. 6 *Cymodocea serrulata* (R.Br) Asch & Magnus

Cymodoceae serrulata is generally referred as 'manatee grass' family or 'serrated seagrass', a dioecious plant, reproduced by rhizome vegetatively. Male flower is located inside the leaf sheath, while the female flower occurs in pairs with elongated stigma found at the base of the leaves. The seeds are rhizome attached, dark and hard, three ridges were found in lengthwise. On low tides, the seeds release and bind to stigma when the tide is high (Menez *et al.*, 1983; Cox,1993).

Fig. 7a. Collection of Seagrasses from mandabam coastal region, Ramanathapuram.

Fig. 7b. Habitat of seagrasses   Fig. 7c. Shade dried seagrasses

4.1.2. Extraction:

The dried materials were crushed in an electric machine blender to make into a coarse powder (Fig. 8; Fig 9 a-e), sequential extraction was done by soaking each seagrass powder in the three analytical grade solvents in the increasing order of polarity such as hexane, Dichloromethane (DCM) (Rankem, India) and Ethanol (Thermofischer's scientific company, India) in the ratio (1:3 w/v powder: solvent) incubated for 48 hours at room temperature (RT) (Fig.10 a-e), they were filtered using 1-2 layers of filter paper (pore size 1µm) placed on Buchner's porcelain filter funnel of 120 mm diameter. The filtrate collected was concentrated at 50° C in the rotary evaporator (Fig. 11 a-c) (Equitron model cat:8925; REG.760; Vacuum Regulator-Analogue, The science House, Chennai, India). To produce a better yield of seagrass crude extracts, the above mentioned procedure was followed twice, the concentrated extracts were thoroughly dried through rotary vacuum pressure of 600 atm/psi and stored in a clean glass bottles for further studies (Fig. 12).

Fig. 8 Shade dried seagrasses blended into coarse powder

Fig. 9a-e Coarse powder of seagrasses
9a. *H. uninervis*; 9b. *H. ovalis*; 9c. *S. isoetifolium*; 9d. *C. serrulata*; 9e. *E. acoroides*

Fig. 10a. *H. uninervis;* Fig. 10b. *H. ovalis*; Fig. 10c. *S. isoetifolium*; Fig. 10d. *C. serrulata*; Fig. 10e. *E. acoroides* powder soaked in hexane, Dichloromethane (DCM) and Ethanol for 48 hours at RT

Fig. 11(a-c) Solvent extraction of seagrasses powder using rotary evaporator

36

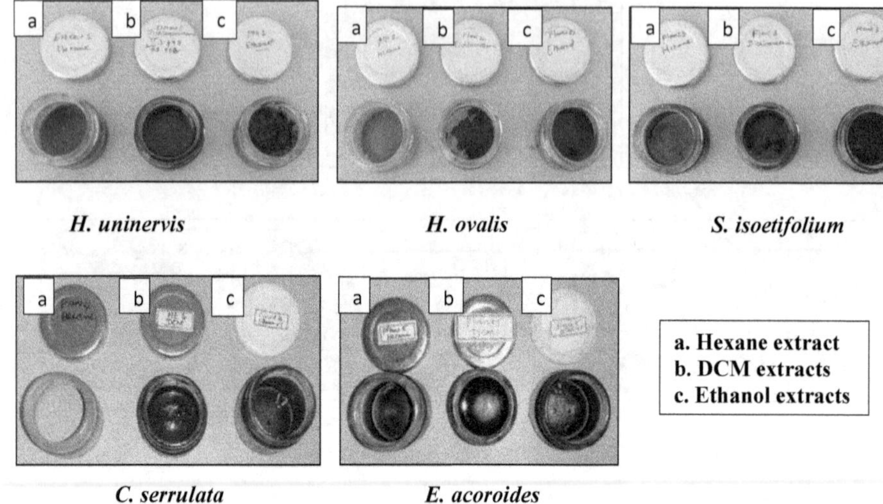

H. uninervis   *H. ovalis*   *S. isoetifolium*

C. serrulata   *E. acoroides*

Fig. 12 Crude hexane, DCM and ethanol solvent extracts of five seagrasses

4.1.3 Preliminary phytochemical analysis:

The extracts of five seagrass species were qualitatively analysed by following the protocols of Senthil Kumar and Reetha, 2009; Naser *et al.*, 2009. About 15 mg of crude extracts of seagrass dissolved in 15 ml of DMSO was used for analysis and DMSO was used as standard. The procedure for the analysis follows below:

Test for Aminoacids:

To 1 ml of sample, 2 drops of ninhydrin was added, purple colour indicates the presence of aminoacids.

Test for Protein:

To 1 ml of sample, 0.5ml of dilute NaoH and 0.5ml of copper sulphate was added, deep blue colour indicates presence of protein.

Test for Tannins:

To 1 ml of sample, few drops of lead acetate in water was added, brown or yellow brown or flocculent precipitate indicate the presence of tannins.

Test for saponins:

1 ml of sample is mixed with 1 ml of water, allow it to vortex in the cyclomixer for two minutes, permanent lather appear on the top layer indicates the presence of saponins.

Test for Flavonoids:

To 1 ml of sample few drops of 2N NaoH was added, yellow colour indicates the presence of flavonoids.

Test for Alkaloids:

To 1 ml of sample, few drops of dilute HCl was added, presence of green colour or dark yellow or brown indicates the presence of alkaloids.

Test for Quinones:

To 1 ml of sample, few drops of concentrated H_2SO_4 was added, reddish pink colour or pink colour indicates the presence of quinones.

Test for Glycosides:

To 1 ml of sample, few drops of alcoholic NaoH was added, pink or red colour indicates the presence of glycosides.

Test for Triterpenoids:

To 1 ml of sample, two drops of acetic anhydride and 1 drop of concentrated H_2SO_4 was added, blue green colour indicates the presence of triterpenoids.

Test for Phenols:

To 1 ml of sample, a few drops of 0.5% ferric chloride in alcohol was added, blue green colour indicates presence of phenols.

Test for steroids:

To 1 ml of sample, a few drops of acetic anhydride and concentrated H_2SO_4 was added, appearance of brown ring indicates the presence of steroids.

Test for Coumarin:

To 1 ml of substance, a few drops of alcoholic NaoH was added, yellow colour, indicates the presence of coumarin.

4.2. Mosquito culture, larvicidal activity of crude extract and statistical analysis:

4.2.1. Mosquito larvae culture:

Mixed stages of *C. quinquefasciatus* (larvae, eggs and pupa) were collected randomly from Cooum river water in the latitude - 13°7'41 North; longitude - 80 °15'58" East (Fig.13a; Fig.13b) and

Ae. aegypti larvae were collected from open clear stored water in the latitude - 13°7'40 North; longitude – 80 °15'52" East (Fig 14a; Fig.14b) North Chennai, Kodungaiyur, Tondairpet Zone, Tamil Nadu, India. They were transported to the lab, segregated by morphological identification and movement and the species authentication was done by Dr. M. Gabriel Paulraj, Senior scientist, Entomology Research Institute, Loyola College, Chennai-34. The pathogen virus free larvae were raised in ceramic bowls (25×35×10 cm) using tap water. A pinch of dog biscuit and brewer's yeast in the ratio of 3:1 was supplemented to larvae for the growth and development. The pupae were collected, transferred into mosquito cages, to emerge in to adult. Periodic blood meal was provided for the adults to deposit eggs (Arivoli and Samuel, 2011). To increase the survival rate of adult mosquitoes, they were feed with 10% sucrose solution and dry grapes. The culture room was maintained at 27 ± 1° C for *Ae. aegypti* and 29 ± 2° C for *C. quinquefasciatus*; photoperiod 11 ± 1 hour; 65-70% rh respectively. The culture is maintained by following the above procedures (Fig.15). Usually third instar larvae was selected for conducting larvicidal screening (Rahman *et al.*, 2000; Mahyoub *et al.*, 2016)

Fig. 13a Collection of random stages of *C. quinquefasciatus* from cooum river water, North Chennai, Kodungaiyur, Tondairpet Zone, Tamil Nadu, India.
Fig. 13b GPS map location of *C. quinquefasciatus* collection.

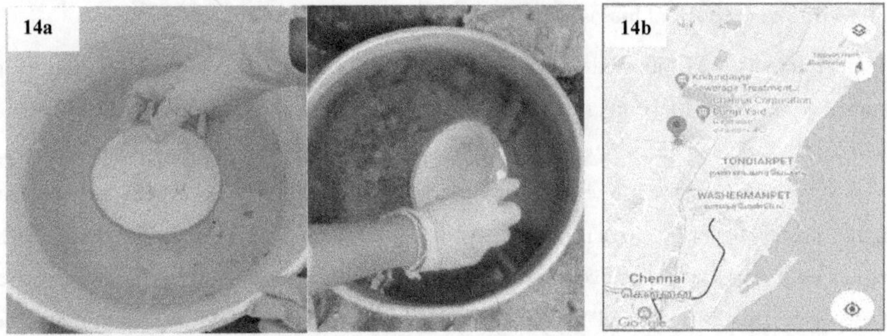

Fig. 14a: Collection of larvae of *Ae. aegypti*, in open stored waters, North Chennai, Kodungaiyur, Tamil Nadu, India.
Fig. 14b: GPS map location of *Ae. aegypti* collection

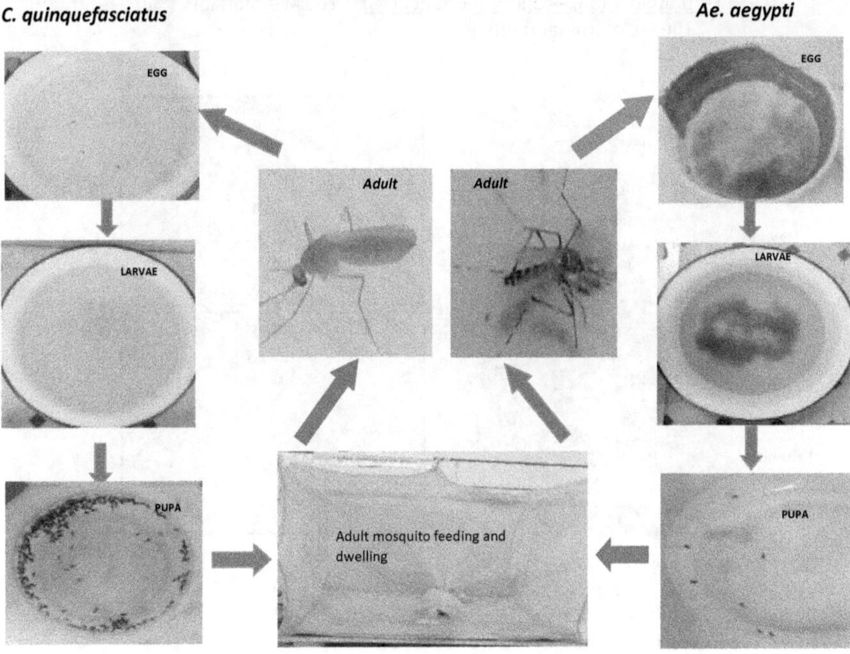

Fig. 15: Rearing of *C. quinquefasciatus* and *Ae. aegypti* larvae in laboratorial conditions

4.2.2. Mosquito Larvicidal bioassay for crude extracts of seagrasses:

Larvicidal activity was followed by the method of WHO, 2005 with slight modifications of Rahuman *et al.*, 2000; Mayoub *et al.*, 2016. Twenty laboratorically reared third instar larvae of *C. quinquefasciatus* and *Ae. aegypti* (Fig. 16b; Fig. 16c) was introduced into each container having 100 ml of untreated water dissolved in different concentrations of seagrass extracts such as 62.5, 125, 250 and 500 ppm (Fig. 16a) using 400 µl of DMSO including its five replicates (Fig.17). The untreated water was kept as positive control and DMSO as negative control. The mortality of larvae was evaluated after 24 hours and 48 hours exposure, they were considered as dead, if it does not move when shaken with the needle. The data were tabulated and the death rate was calculated using abbott's formula (Abbott's Formula, 1925).

$$\frac{\text{Test mortality (\%) - control mortality (\%)}}{100 - \text{Control mortality (\%)}} \times 100\% = \text{mortality (\%)}$$

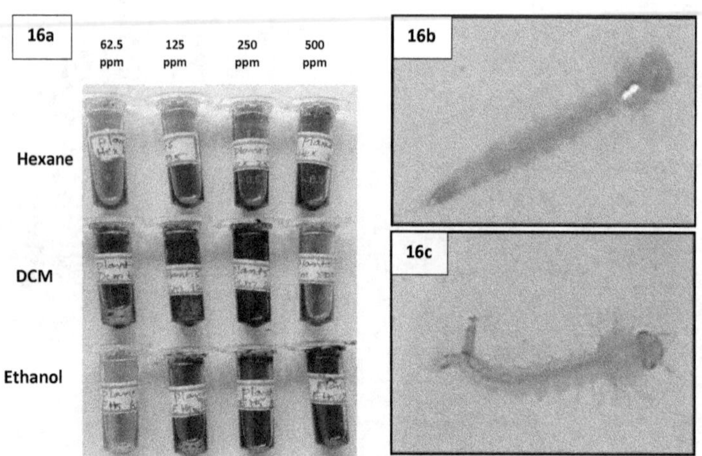

Fig. 16a Preparation of solvent extraction (62.5-500 ppm)
Fig. 16b Third instar larvae of *Ae. aegypti*
Fig. 16c Third instar larvae of *C. quinquefasciatus*

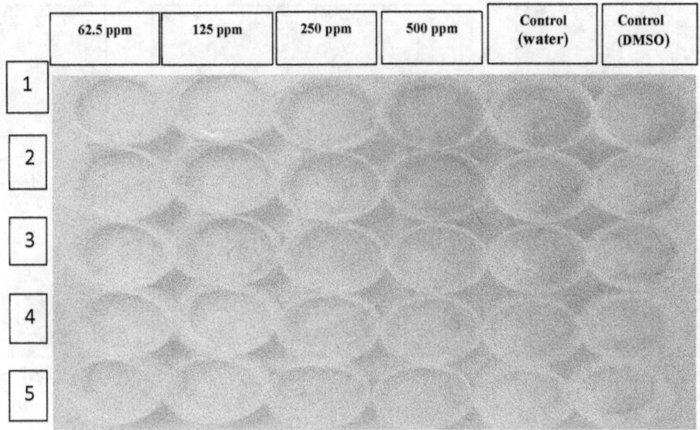

Fig. 17 Experimental setup of larvicidal bioassay showing 5 replicates with respective positive control (water) and negative control (DMSO)

4.2.3. Statistical analysis:

The data were subjected to statistical analysis using SPSS (Statistical package for social sciences software) version 21.0. The values of dose concentration, total number of larvae introduced, responded larvae were inputted using Probit analysis (Finney,1971). The results of LC_{50} (Concentration of sample that kills 50 % organisms) and LC_{90} (Concentration of sample that kills 90 % organisms) upper and lower fiducial confidence limits, Slope concept, Correlation coefficient (r), standard deviation (SD), Chi square test (χ^2) were calculated.

4.2.4. Larvicidal activity of hexane extracts of epiphytic and non-epiphytic portion of *C. serrulata* leaves against third instar larvae of *C. quinquefasciatus*:

From the above analysis, crude hexane extracts of *C. Serrulata* leaves was found efficient against third instar larvae of *C. quinquefasciatus* and *Ae. aegypti* but from the literature investigation, seagrasses are influenced with some epiphytes (Takahashi, 2016). This was taken for further research by analysing the larvicidal efficacy of epiphytes and non-epiphytic portion (Only leaf portions) separately (Fig.18; Fig.19) by implementing the above methodology.

  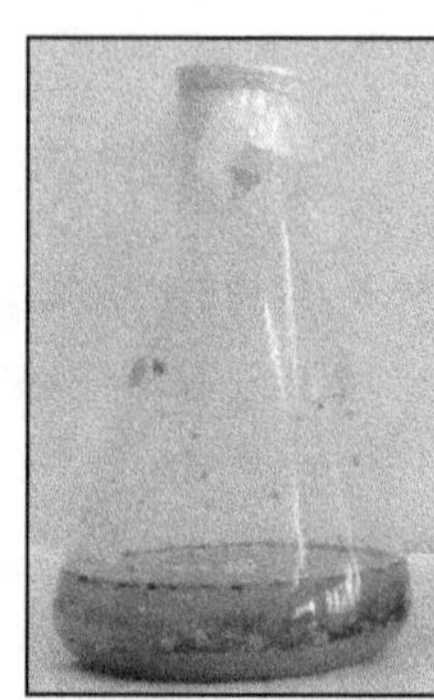

| **Epiphytic leaves of *C. serrulata*** | **Epiphytes** | **Soaked in hexane for 48 hours** |

Fig. 18 Processing of epiphytes of *C. serrulata* leaves for extraction

  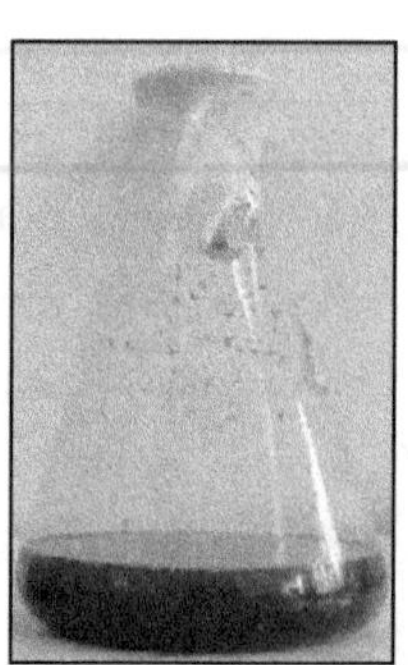

| **Non-Epiphytic leaves of *C. serrulata*** | **Coarse powder** | **Soaked in hexane for 48 hours** |

Fig.19: Processing of non-epiphytic portion (only leaves) of *C. serrulata* for extraction

4.3: Isolation and larvicidal activity of active fraction of *C. serrulata* leaves:

4.3.1. Bioassay guided fractionation by column chromatography:

To isolate the active compound, column with a diameter 2.2 cm and a length of 32.5 cm was chosen, washed twice with acetone and then rinsed with the hexane before being allowed to evaporate overnight. With the help of a long glass rod, a small amount of cotton was filled in the bottom of the column. Silica gel of 100-200 mesh was mixed with the hexane in the ratio (1:3), gradually poured and allowed it to settle 3/4th level of the way down in to the column. Once the silica gel settled, the effective larvicidal hexane extract of the *C. serrulata* was progressively introduced into the column then it was closed with Cap/aluminium foil. The knob of the column was slowly released to elute 20 ml of fractions one after the other by loading the solvent hexane: ethyl acetate onto the column in varied ratios (9.5:0.5, 9:1; 4:1; 4:2 and 4:1). All the fractions were concentrated and analysed using thin layer chromatography.

4.3.2. Thin Layer Chromatography (TLC):

One of the most fundamental and effective technique for separating compounds is done through thin layer chromatography. This was carried out using commercially available aluminium plates coated with silica gel considered as stationary phase, while the solvents hexane: ethyl acetate used as mobile phase. The plates were trimmed in to small size (2×5 cm) and the concentrated fractions (solute) were loaded one at a time and allow it to air dry. The plates loaded with solutes were placed in coupling jar standardised with hexane and ethyl acetate in the ratios of 4.5:0.5 and 4:1, respectively and allow to run. The resolution factor 'R_f' was calculated using the formula:

$$R_f \quad = \quad \frac{\text{Distance travelled by solute front}}{\text{Distance travelled by solvent front}}$$

4.3.3 Larvicidal activity of fractions:

From the TLC analysis, eight fractions were found and named as A to H, each fraction was subjected for larvicidal activity with 2.5, 5 and 10 ppm concentrations against third instar larvae of *C. quinquefasciatus* and *Ae. aegypti* including five replicates.

4.3.4 Isolation of subfractions:

From the analysis, Fraction E was found potent against third instar larvae of *C. quinquefasciatus* and *Ae. aegypti,* with the reference to 'R_f' values of TLC analysis, two prominent bands were identified (Compound 'A' and Compound 'B'). Further, they were separated through small size column chromatography of 1.8 cm diameter and 32.5 cm length, the compounds were eluted using the solvents hexane: ethyl acetate in the ratio 9:1 and 8:2.

4.3.5. larvicidal activity of compound 'A' and 'B':

The larvicidal activity was carried out by slightly reducing the concentrations of compound 'A' and compound 'B' to 1, 2.5 and 5 ppm with its five replicates along with untreated water as positive control, azadirachtin (Natural compound) and Temephos (Synthetic Compound) as negative control were tested against third instar larvae of *C. quinquefasciatus* and *Ae. aegypti* .

4.4. Structural Characterisation of compound 'A' of *C. serrulata* leaves

4.4.1. UV-Vis spectroscopy:

UV-Vis spectroscopy was performed in JASCO V - 670 Spectrophotometer. The spectral readings are scanned and saved.

4.4.2. Fourier Transform Infrared Spectroscopy (FTIR):

About 0.2 mg of dried extract of compound 'A' dissolved in 1 ml of chloroform to detect the various frequencies of different functional groups. The spectral study was accomplished by the FTIR instrument model IRAffinity-1S by ATR mode was supplied by Shimadzu, Japan with the wave number ranging from 3430-840 cm^{-1}.

4.4.3. Proton-Nuclear Magnetic Resonance (^{1}H NMR):

The ^{1}H NMR spectra analysis was performed in Bruker instrument, resonates at 400 MHz using CDCl$_3$. Chemical shift values were given in delta scale and tetramethylsilane is used as internal standard.

4.4.4. Gas Chromatography-Mass Spectrometer (GC-MS):

GC-MS analysis of compound 'A' was carried out using Clarus 680 GC model. It is made up of gas chromatogram integrated with the mass spectrometer connected to the fused silica column, packed with Elite-5MS (5% biphenyl 95% dimethylpolysiloxane, 30 m × 0.25 mm ID × 250 μm df) and mass detector operated at 230°C and ionization energy of 70 eV was utilised. The components were separated using helium gas (as a carrier) at a constant flow of 1 ml/min. The extract sample of 1 μL was injected into the instrument at the oven temperature of 60 °C for 2 min and then increased to 300 °C at the rate of 10 °C/min and the injector temperature was set at 260°C during the chromatographic run. The spectral data were compared with the database of known components which are stored in the GCMS NIST library from the software turbo Mass version.

4.5: Histopathological alterations of third instar larvae of *C. quinquefasciatus* against isolated compound 'A' of *C. Serrulata* leaves:

Third instar larvae of *C. quinquefasciatus* treated with Compound 'A', temephos and untreated larvae (tap water) were encountered histopathological studies. According to the method done by Raymond *et al.*, 2007 with slight modification, the tissues of larvae were fixed in 1% formalin for 24 hours. The tissues are dehydrated using 50, 60, 70, 80, 90 and 100% alcohol every two hours. After dehydration, it is placed in xylene for 5 hours and then transferred into warm oven with wax for embedding. The liquid wax was poured into paper blocks, allow to cool and used for sections.

Using microtome sectioning was done and they were placed in slide, again dewaxed with xylene and alcohol from 100% to 50% then, washed with distilled water and stained with haematoxylin. They were again dehydrated in alcoholic series and then stained in eosin. Finally, it was washed in 100% alcohol, dipped into xylene twice and mounted in DPX. The disrupted mid gut larval cells of untreated, compound 'A' and temephos were photographed in high resolution microscope connected to computer.

4.6. Non-target organisms of *Poecilia reticulata* (Guppy fish) and *Gambusia affinis* (Mosquito fish) against compound 'A' of *C. serrulata* leaves:

The isolated compound 'A' was further tested against non-target organisms such as predatory fish *Poecilia reticulata* and *Gambusia affinis* collected from Fishery Research Institute, Chetpet, India, with the slight modification analysed by Ignachimuthu and Maheswaran (2012). The compound 'A' and its standard positive controls (azadiractin and temephos) was prepared into four different concentrations 0.5, 1, 2.5 and 5 ppm, including its five replicates dissolved in 400 µl of DMSO mixed in container containing 100ml of tap water. About 10 predatory fish were introduced into each container. The abnormalities and mortality will be observed after 24 hours exposure. The Suitability index factor (SIF) or Predatory safety factor (PSF) was calculated by finding the LC_{50} and LC_{90} values from Probit analysis, illustrated by Deo and Colleagues, 1998

$$\text{SIF/PSF} = \frac{LC_{50} \text{ of non-target of organisms exposed to isolated compound}}{LC_{50} \text{ of target vector species exposed to isolated compound}}$$

4.7: *In silico* molecular docking analysis of *ace1* receptor of *C. quinquefasciatus* and *Ae. aegypti* against compounds 'A' of *C. serrulata* leaves

4.7.1. Ligand Preparation:

The compound 'A' (Ligand) responsible for larvicidal activity predicted through GC-MS analysis and NMR analysis was found as stigmasterol, campesterol and betasitosterol. To find their individual efficacy, 2-Dimentional structure retrieved from PubChem database (http://pubchem.ncbi.nlm.nih.gov/) was converted to 3D structure using ChemSketch software (Fig. 20a; Fig. 20b; Fig. 20c) along with control azadirachtin (Fig. 20d) and temephos (Fig. 20e), were saved in Mol2 format.

4.7.2. Receptor Preparation:

The template sequence of *ace1* receptor (Id: Q867X2) of *C. quinquefasciatus* and *Ae. aegypti* (Id: Q6A2G6) was found in UNIPROT database (https://www.uniprot.org) . They were BLAST to find 100% similarly matched sequence. The matched sequence was downloaded in FASTA format and paste it in the SWISS MODEL to build a homologous protein model. The best matched identitical sequence for *Ae. aegypti* and *C. quinquefasciatus* found as 92.78% and 93.94% and the most highlighted homologous models were 5ydi.2.A and 5x61.1, downloaded in PDB format and visualised in UCSF Chimera 1.14 version. Then the polar bonds, amino acids, active site and distance between the hydrogen atoms were selected (Fig. 21; Fig. 22)

4.7.3. Ramachandran Plot:

Ramachandran plot was analysed to find the structural quality, chemical quality and the favourable regions of aminoacids of the predicted model (Fig. 23; Fig. 24) (Fournier and Mutero, 1994)

4.7.4. Molecular Docking:

Molecular docking was performed using online software SWISSDOCK by uploading 3D structures of ligands including the controls in Mol2 format and *ace1* receptor of *C. quinquefasciatus* and *Ae. aegypti* in PDB format. Auto grid calculations was automatically performed during the run, different binding modes were executed but, the most compatible receptor–ligand complex was selected and downloaded. The binding energy (ΔG) and binding interaction (Fullness) Kcal/Mol was evaluated by cluster scoring (Grosdidier *et al.*, 2007).

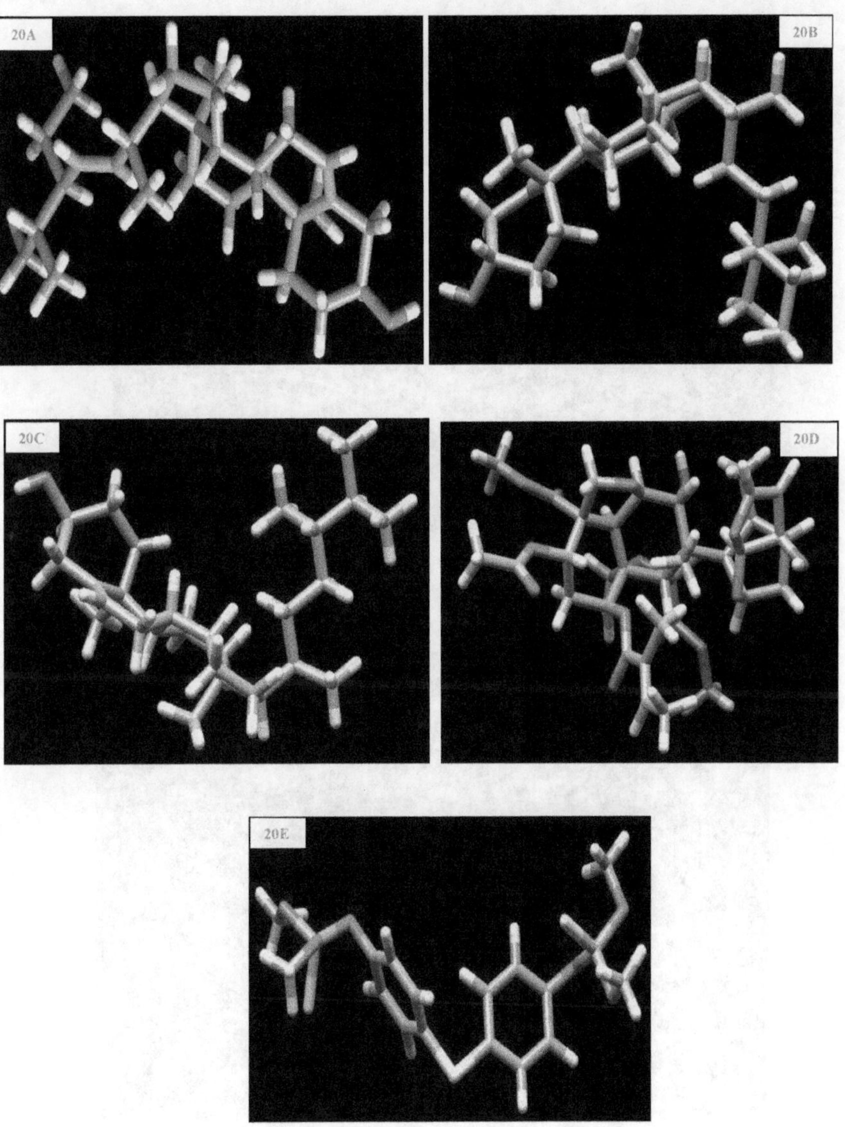

Fig 20: Three -Dimentional Ligand structures – Stigmasterol (20A); Campesterol (20B); Bitasitosterol (20C); Azadirachtin (20D); Temephos (20D)

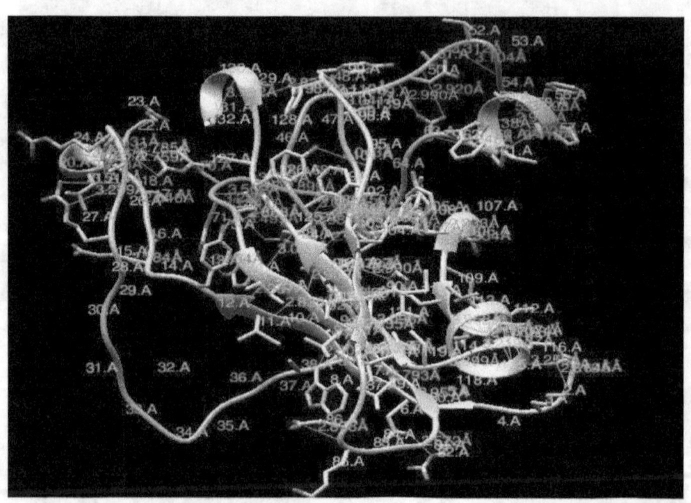

Fig 21 Homology model of *ace1* receptor of *C. quinquefasciatus* showing hydrogen bonds and active amino acids

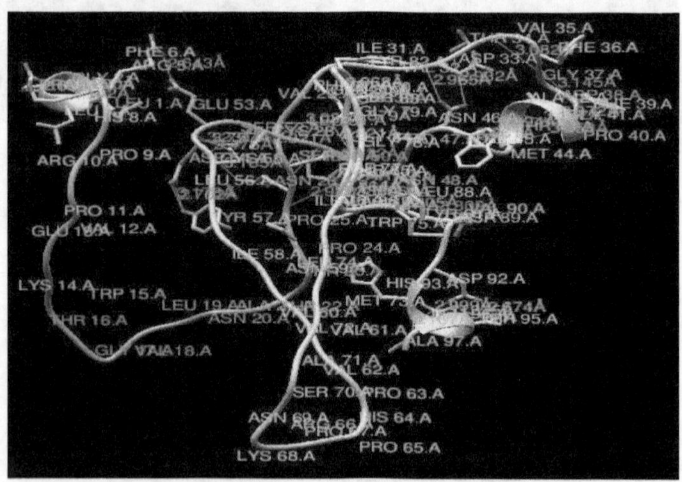

Fig. 22 Homology model of *ace1* receptor of *Ae. aegypti* showing hydrogen bonds and active amino acids

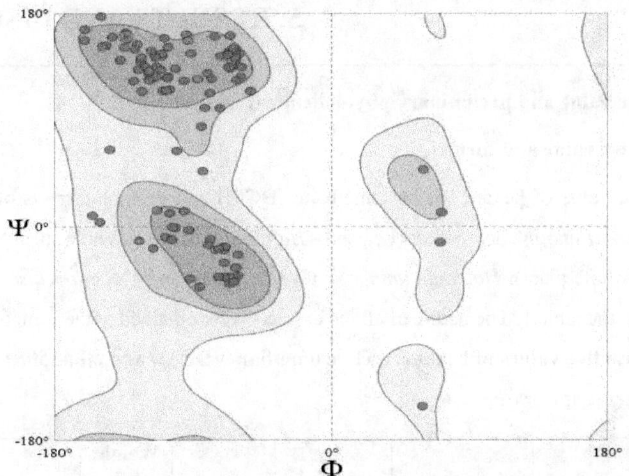

Fig. 23 Ramachandran Plot for homology model of *ace1* receptor
C. quinquefasciatus

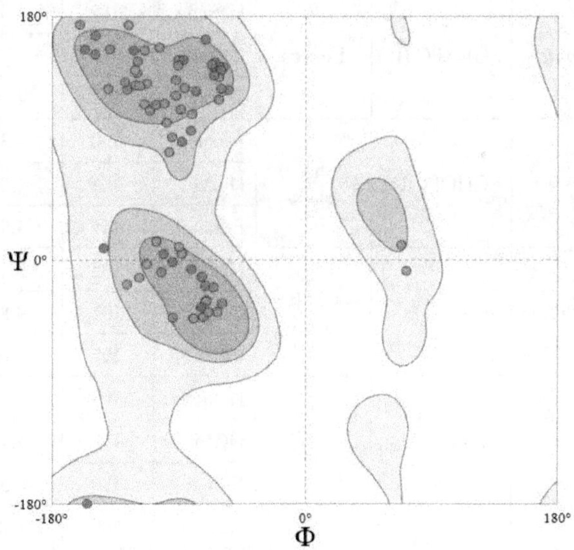

Fig. 24 Ramachandran Plot for homology model of *ace1* receptor of *Ae. aegypti*

5.1 Extractive value and preliminary phytochemical analysis:

5.1.1 Extractive value and form:

The extractive value of hexane, dichloromethane (DCM) and ethanol extracts of five different seagrasses *viz., Enhalus acoroides* (Leaves), *Halophila ovalis* (Whole plant), *Syringodium isoetifolium* (whole plant), *Halodule uninervis* (Whole plant) and *Cymodocea serrulata* (leaves) were shown in the table.1. The nature of all the extracts were obtained in the form of solid and gel.

Table 1: Extractive values of hexane, dichloromethane (DCM) and ethanolic extracts of five seagrass species

S.No.	Sea grasses	Voucher Number	Parts used	Solvent Extracts	Weight of the crude powder (g)	Extractive value(g)	Extract form
1	*E. acoroides*	SGEAPC01	Leaves	Hexane	615	3.336	Solid
				DCM	611	3.801	Solid
				Ethanol	507	4.029	Solid
2	*H. ovalis*	SGHOPC02	whole plant	Hexane	450	3.376	Solid
				DCM	445	4.786	Solid
				Ethanol	441	4.977	Gel
3	*S. isoetifolium*	SGSIPC03	whole plant	Hexane	305	3.322	Gel
				DCM	301	3.879	Gel
				Ethanol	297	4.038	Solid
4	*H. uninervis*	SGHUPC04	whole plant	Hexane	505	6.738	Solid
				DCM	498	7.132	Gel
				Ethanol	490	8.588	Solid
5	*C. serrulata*	SGCSPC05	Leaves	Hexane	600	3.826	Solid
				DCM	596	4.641	Gel
				Ethanol	591	4.137	Solid

5.1.2 Preliminary phytochemical analysis:

Preliminary phytochemical analysis of crude extracts of five different seagrasses are compacted in table.2; figures 25-29. According to the present findings, *H. uninervis* showed the presence of tannins, quinones, phenols in ethanolic extracts; triterpenoids was found positive in hexane extracts. The analysis of *H. ovalis* showed the presence of tannins, flavonoids, phenol, coumarin in ethanolic extracts; presence of flavonoids, alkaloids, triterpenoids and phenol in the DCM extracts; saponins were detected in hexane extracts. The analysis from *S. isoetifolium* showed presence of flavonoids, triterpenoids and phenol in DCM extracts whereas presence of saponin, alkaloid and glycosides in hexane extracts. The analysis of *C. serrulata* showed the presence of steroids in hexane extracts; tannin in DCM and ethanol extracts whereas flavonoids, alkaloids and triterpenoids were found only in DCM extracts. Analysis from *E. acoroides* showed the presence of steroids in hexane extracts; tannins was observed in DCM and ethanolic extracts; presence of phenol found only in DCM extracts.

In the preceding studies, Ravikumar *et al.*, 2011c disclosed phenol, triterpenoids, flavonoids, saponins and alkaloids isolated from marine plant extracts are effective as single or synergistic compounds, which influences the mosquito larvicidal action. Filho *et al.*, 2009 reported reducing sugars and phenolic groups has potential mosquito larvicide. Vijayakumar *et al.*, 2014; Fernandes *et al.*, 2019 and Teles *et al.*, 2018 expressed flavonoid sulfate has the ability to inhibit the mosquito larvae alterations. Cania and Setyaningrum, 2013 proclaimed that alkaloids inhibit the acetylcholinesterase activity of the mosquito larvae. Moreover, Vijayakumar *et al.*, 2014 observed the presence of quinones, phenol, alkaloids and absence of steroids from the crude extracts of *C. serrulata* but in the present study it is significantly answered limbermann Burchard test, indicating the presence of steroids.

Table. 2 Preliminary phytochemical analysis of hexane, dichloromethane and ethanol extracts of five selected seagrass species

S.No	Phytochemical test	*H.uninervis*			*H.ovalis*			*S.isoetifolium*			*C.serrulata*			*E. acoroides*		
		Hexane	DCM	Ethanol	Hexane	DCM	Ethanol	Hexane	DCM	Ethanol	Hexane	DCM	Ethanol	Hexane	DCM	Ethanol
1	Aminoacid	-	-	-	-	-	-	-	-	-	-	-	-	-	-	-
2	Protein	-	-	-	-	-	-	-	-	-	-	-	-	-	-	-
3	Tannins	-	-	+	-	-	+	-	-	-	-	+	+	-	+	+
4	Saponins	-	-	-	+	-	-	-	-	+	-	-	-	-	-	-
5	Flavonoids	-	-	-	-	+	+	-	+	-	-	+	-	-	-	-
6	Alkaloids	-	-	-	-	+	-	-	-	+	-	+	-	-	-	-
7	Quinones	-	-	+	-	-	-	-	-	-	-	-	-	-	-	-
8	Glycosides	-	-	-	-	-	-	-	-	+	-	-	-	-	-	-
9	Triperpenoids	+	-	-	-	+	-	-	+	-	-	+	-	-	-	-
10	Phenol Test	-	-	+	-	+	+	-	+	-	-	-	-	-	+	-
11	Steroids	-	-	-	-	-	-	-	-	-	+	-	-	-	-	-
12	Coumarin	-	-	-	-	-	+	-	-	-	-	-	-	-	-	-

'+' Presence '−' absence;

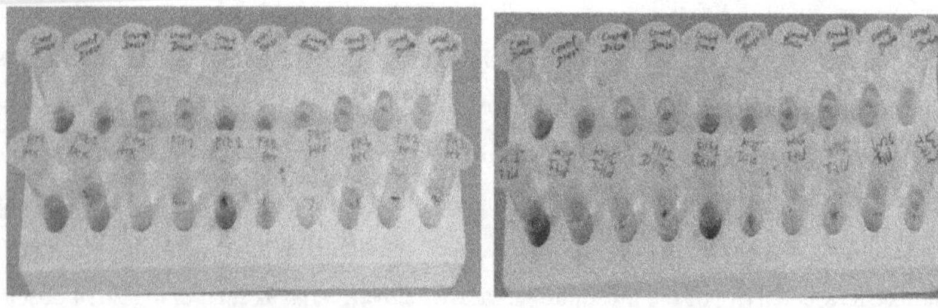

Hexane extract   **Dichloromethane extract**

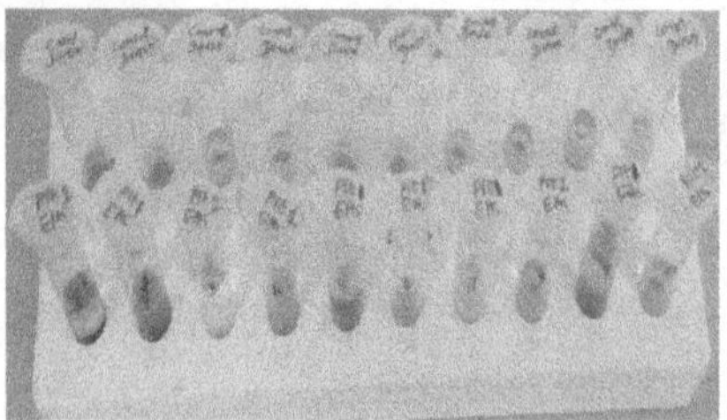

Ethanol extract

Fig 25 Preliminary phytochemical analysis of whole plant of *H. uninervis* extracts

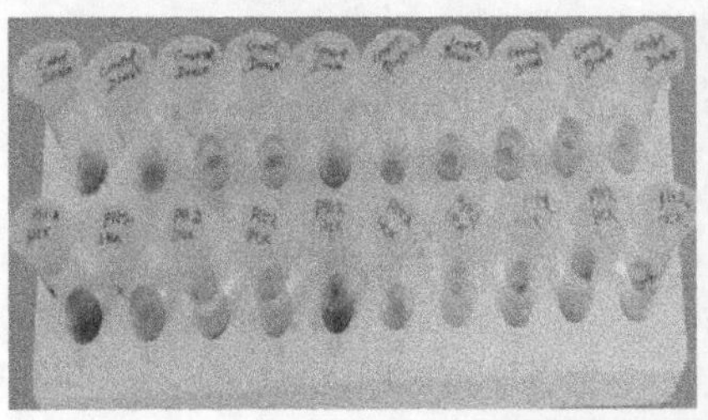

Hexane extract

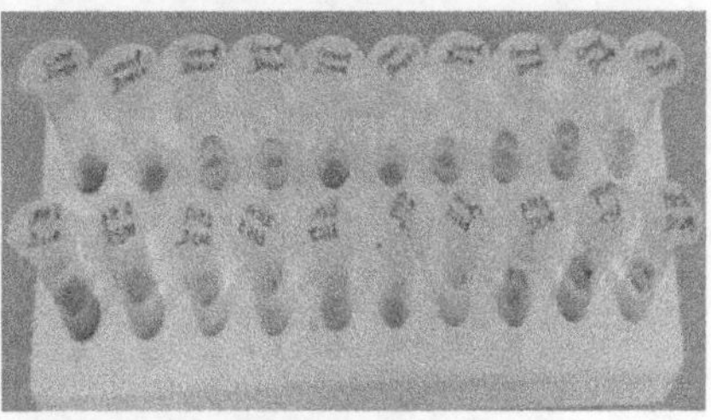

Dichloromethane extract

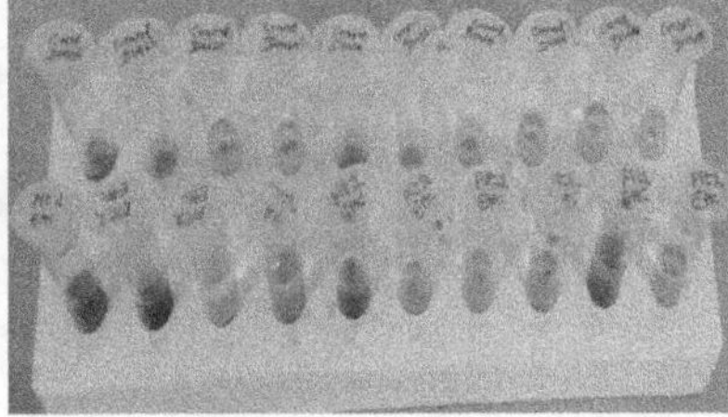

Ethanol extract

Fig. 26 Preliminary phytochemical analysis of whole plant extracts of *H. ovalis*

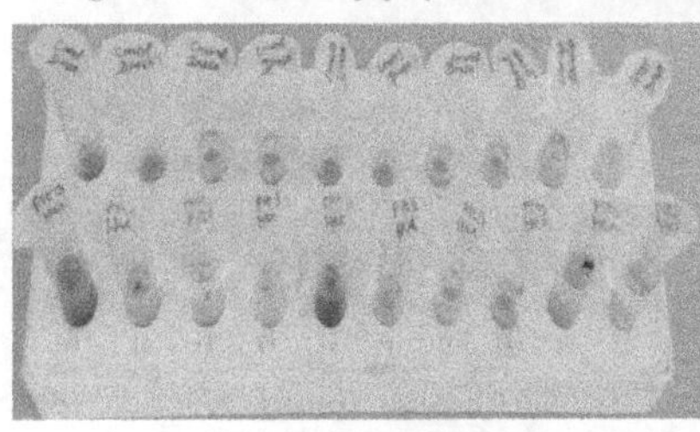

Hexane extract

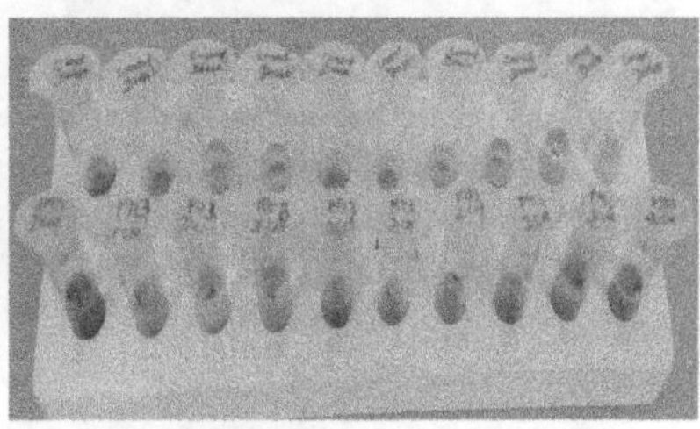

Dichloromethane exttract

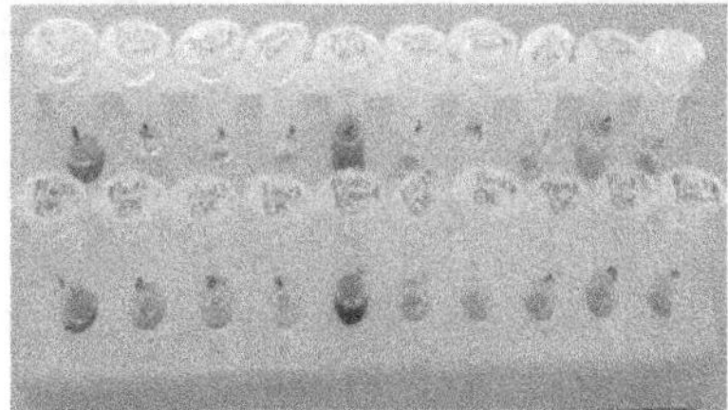

Ethanol extract

Fig 27 Preliminary phytochemical analysis of whole plant extract of *S. isoetifolium*

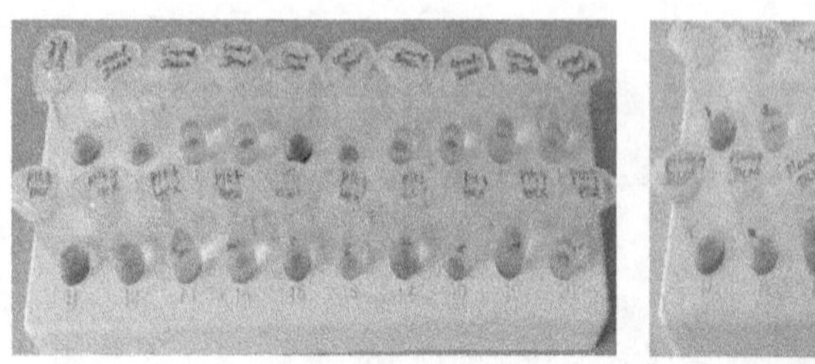

Hexane extract

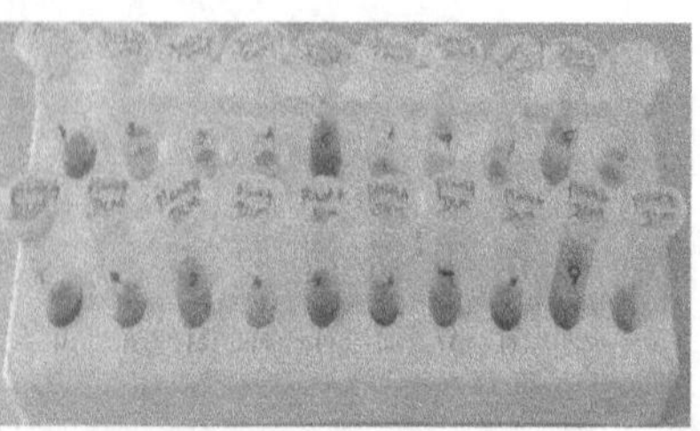

Dichloromethane extract

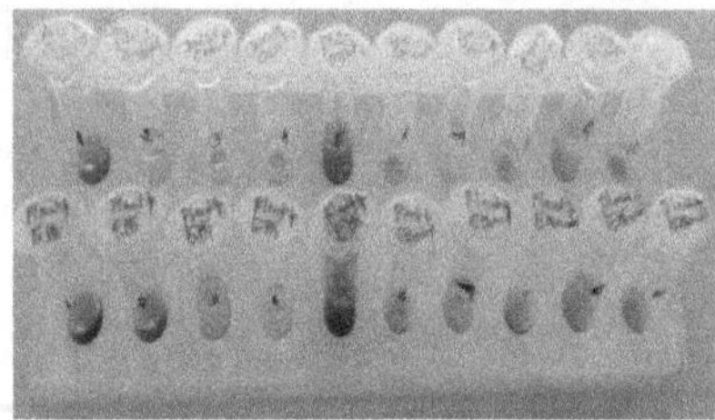

Ethanol extract

Fig 28 Preliminary phytochemical analysis of *Cymodocea serrulata* leaf extracts

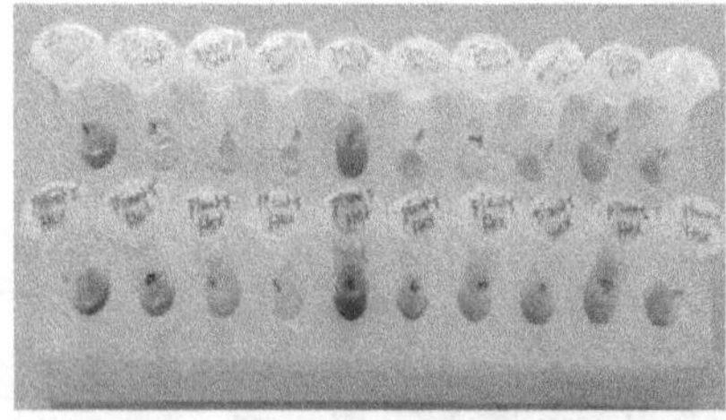

Hexane extract

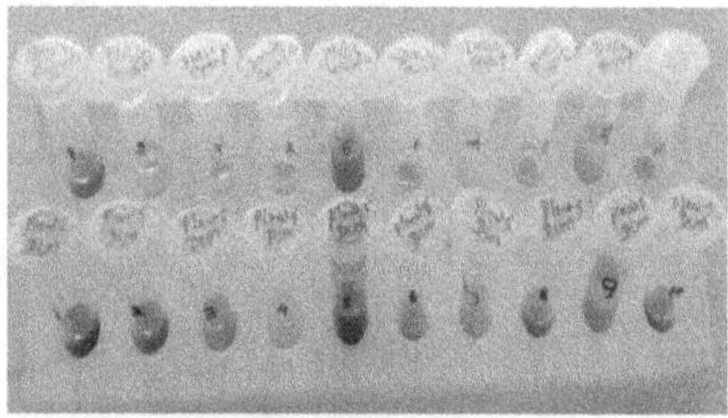

Dichloromethane extract

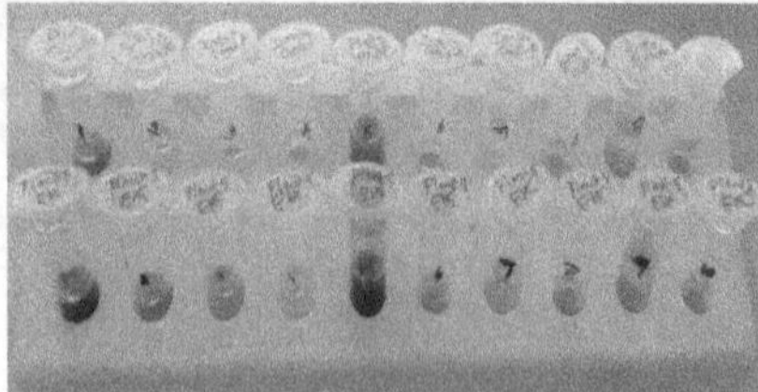

Ethanol extract

Fig 29 Preliminary phytochemical activity of *Enhalus acoroides* leaf extracts

5.2. Larvicidal activty of crude, epiphytes and non-epiphytic solvent extracts:

5.2.1. Larvicidal activity of crude solvent extracts of five seagrass species:

Graph.1 and 2; tables.3 and 4 shows the larvicidal activity of crude hexane, DCM and ethanolic extracts of five species of seagrass species against third instar larvae of *C. quinquefasciatus* at four different concentrations *viz.*, 62.5, 125, 250 and 500 ppm. The maximum mortality of 90 % and 94 % was recorded in hexane extracts of *C. serrulata* leaves against *C. quinquefasciatus* with the LC_{50} values of 0.022 and 0.129 µg/ml; LC_{90} values of 0.336 and 0.817 µg/ml, respectively. The moderate activity was found in hexane extracts of *H. uninervis* around 48% and 72% with the LC_{50} values of 0.006 and 0.012 µg/ml; LC_{90} values of 0.018 and 0.079 µg/ml, respectively. Followed by, ethanolic extracts of *E. acoroides* leaves was found active at 50% and 54%, with the LC_{50} value of 0.476 and 0.537 g/ml; LC_{90} values of 1.921 and 2.153 µg/ml, respectively after 24 and 48 hours treatments.

Similarly, the effect of hexane, DCM and ethanolic crude solvent extracts of five different seagrasses was tested against third instar larvae of *Ae. aegypti* and the results are represented as graph. 3 and 4; table. 5 and 6. The highest mortality observed as 80 and 86% in the leaf hexane extracts of *C. serrulata* at 500 ppm with the LC_{50} values of 0.006 and 0.030 µg/ml; LC_{90} values of 1.364 and 2.019 µg/ml, respectively. The second highest mortality was found in hexane extracts of *E. acoroides* as 62% and 66% with the LC_{50} values of 0.068 and 0.191 µg/ml, respectively; LC_{90} values of 6.064 and 40.518 µg/ml, respectively. Followed by, moderate activity was reported in hexane extracts of *H. uninervis* as 46% and 74% with the LC_{50} values of 0.226 and 0.474 µg/ml; LC_{90} values of 0.690 and 1.065 µg/ml, respectively. The Minimal activity was showed in hexane extracts of *S. isoetifolium* as 26% and 50% with the LC_{50} values of 0.457 and 3.587 µg/ml; LC_{90} values of 18.335 and 34.898 µg/ml, respectively and the least activity was observed in *H. ovalis* as 22% and 30% with the LC_{50} value of 1.204 and 1.342 µg/ml; LC_{90} value of 7.516 and 20.259 µg/ ml, respectively after 24 hours and 48 hours. The larvicidal activity was also recorded in DCM extracts of five different seagrasses from 62.5-500 ppm and the percentage of mortality was found as 8-32% in *H. uninervis*, 1-14% in *H. ovalis*, 1-16% in *S. isoetifolium*, 8 -32% in *C. serrulata* and 4-30% in *E. acoroides* after 24 and 48 hours treatment and no mortality was observed in ethanolic extracts.

From the above studies it is clearly understood that systematic increase in polarity of solvents are important to isolate a specific active larvicidal compound. In previous research, Mahyoub *et al.*, 2016; Yusniawati *et al.*, 2017; Purnomo *et al.*, 2017 studied larvicidal activity only in polar solvent such as ethanolic and methanolic extracts but, there is no better understanding. In the current research, non polar solvent, hexane play a significant role in isolating a active larvicidal compound against *C. quinquefasciatus* and *Ae. aegypti*. Secondly, in comparison to the incubation period of 21 days under dark condition exhibited by Ali *et al.*, 2012; 2013, Vijayakumar *et al.*, 2014, the current research followed only 48 hours of incubation period at room temperature which is very much less time consuming and the compound might not lead to denaturation.

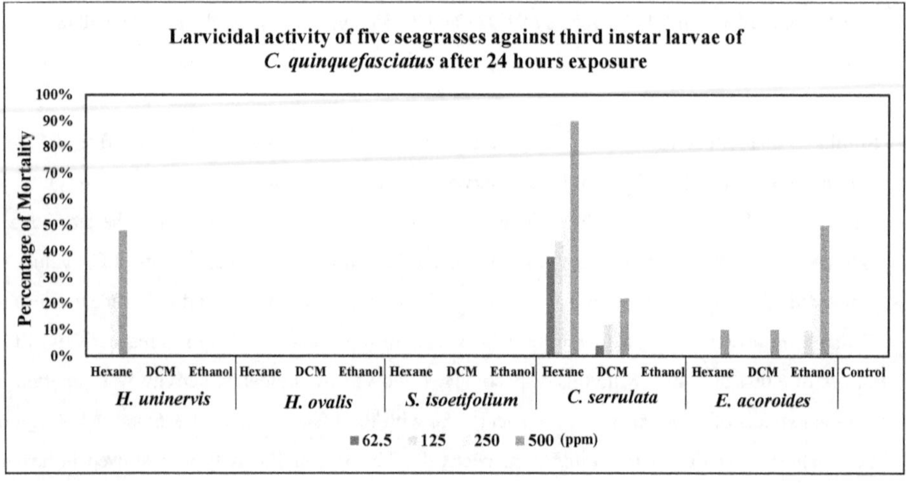

Graph 1 Mosquito larvicidal activity of five seagrasses treated against third instar larvae of *C. quinquefasciatus* after 24 hours exposure

Table 3 Lethal concentrations (µg/ml) of hexane extracts of *H. uninervis*, *C. serrulata* and *E. acoroides* against third instar larvae of *C. quinquefasciatus* after 24 hour post Treatment

Sea grasses	Extracts	LC$_{50}$ LL-UP	LC$_{90}$ LL-UP	Slope±SE	SD	R^2	χ^2
H. uninervis	Hexane	0.006 (0.004-0.009)	0.018 (0.012-0.027)	2.717±0.09	0.368	0.999	0.924
C. serrulata	Hexane	0.022 (0.008-0.060)	0.336 (0.125-0.902)	1.093±0.21	0.915	0.944	0.944
	DCM	1.543 (0.422-5.637)	75.309 (20.614-275.12)	0.765±0.28	1.307	0.859	0.722
E. acoroides	Ethanol	0.476 (0.285-0.796)	1.921 (1.149-3.213)	2.118±0.11	0.472	0.977	0.187

➢ Solvent extracts concentration from 62.5-500 ppm
➢ LL- Lower limit; UP- Upper limit; Correlation coefficient (R), standard deviation (SD), Chi square test (χ^2).

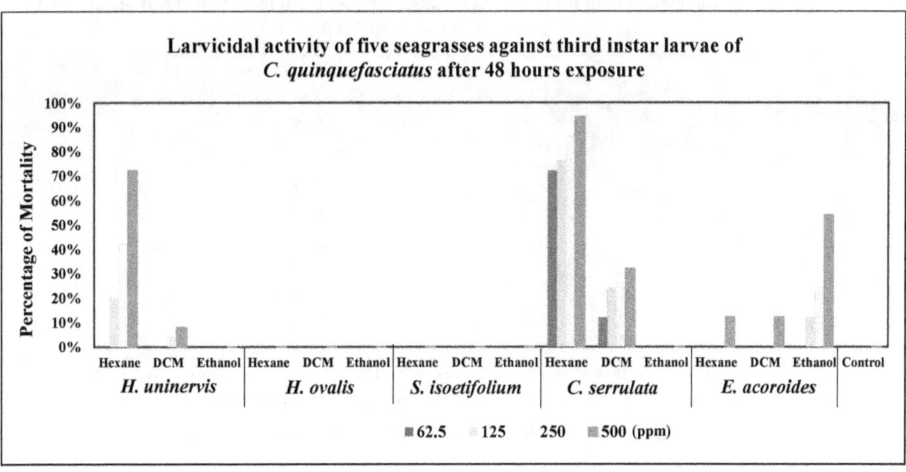

Graph 2 Mosquito larvicidal activity of five seagrasses treated against third instar larvae of *C. quinquefasciatus* after 48 hours exposure

Table 4 **Lethal concentrations (µg/ml) of hexane extracts of *H. uninervis*, *C. serrulata* and *E. acoroides* against third instar *C. quinquefasciatus* larvae after 48 hours post treatment**

Sea grasses	Extracts	LC_{50} LL-UP	LC_{90} LL-UP	Slope±SE	SD	R^2	χ^2
H. uninervis	Hexane	0.012 (0.006-0.024)	0.079 (0.039-0.161)	1.552±0.15	0.644	0.926	0.547
C. serrulata	Hexane	0.129 (0.073-0.230)	0.817 (0.461-1.448)	1.602±0.12	0.601	0.808	0.502
	DCM	2.778 (0.869-8.886)	54.363 (16.998-173.862)	1.007±0.25	0.993	0.904	0.737
E. acoroides	Ethanol	0.537 (0.317-0.909)	2.153 (1.271-3.648)	2.129±0.11	0.470	0.968	0.222

➤ Solvent extracts concentration from 62.5-500 ppm
➤ LL- Lower limit; UP- Upper limit; Correlation coefficient (R), standard deviation (SD), Chi square test (χ^2).

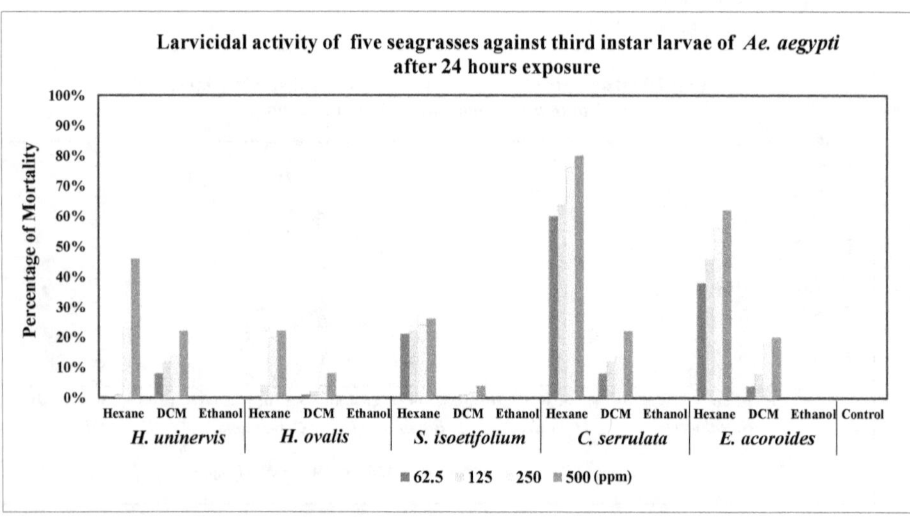

Graph. 3 Comparison graph showing mosquito larvicidal activity of five seagrasses treated against third instar larvae of *Ae. aegypti* after 24 hours exposure

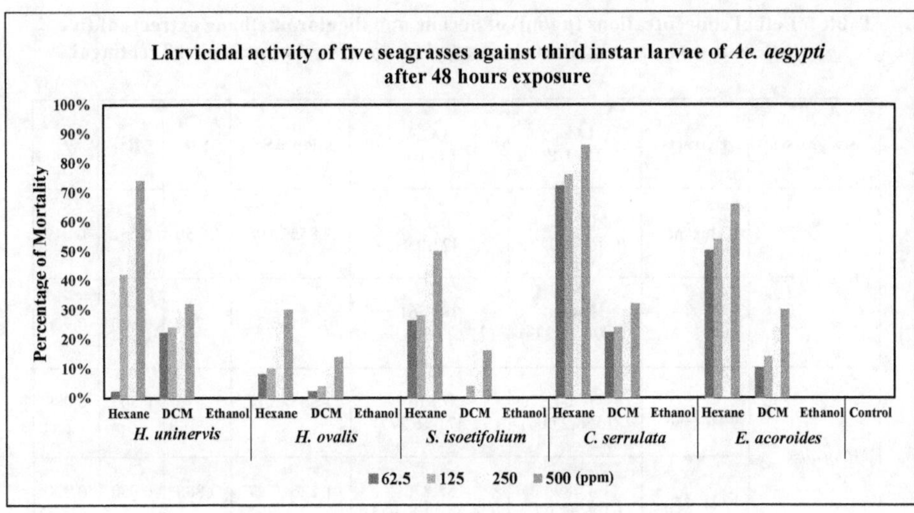

Graph. 4 Comparison graph showing mosquito larvicidal activity of five seagrasses treated against third instar larvae of *Ae. aegypti* after 48 hours exposure

Table 5 Lethal concentrations (µg/ml) of hexane and dichloromethane extracts of five different seagrasses against *Ae. aegypti larvae* after 24 hours post Treatment

Sea grasses	Extracts	LC_{50} LL-UP	LC_{90} LL-UP	Slope±SE	SD	R^2	χ^2
H. uninervis	Hexane	0.226 (0.156-0.327)	0.690 (0.477-0.998)	2.856±0.08	0.350	0.852	0.019
	DCM	2.496 (0.470-13.243)	388.561 (73.221-206.985)	0.585±0.37	1.708	0.920	0.906
H. ovalis	Hexane	1.204 (0.556-2.606)	7.516 (3.471-16.277)	1.625±0.17	0.615	0.803	0.089
	DCM	4.089 (1.196-13.974)	55.872 (16.348-190.945)	1.126±0.27	0.888	0.980	0.920
S. isoetifolium	Hexane	0.457 (0.144-1.449)	18.335 (5.790- 58.059)	0.801±0.25	1.248	0.899	0.821
	DCM	2.924 (0.982-8.707)	30.674 (10.302-91.329)	1.256±0.24	0.796	0.981	0.328
C. serrulata	Hexane	0.006 (0.001-0.037)	1.364 (0.210-8.844)	0.541±0.41	1.849	0.973	0.998
	DCM	1.922 (0.444-8.326)	159.116 (36.725- 689.388)	0.671±0.32	1.490	0.889	0.829
E. acoroides	Hexane	0.068 (0.010-0.482)	6.064 (2.095-17.558)	0.462±0.43	2.165	0.990	0.997
	DCM	1.856 (0.564-6.106)	37.881 (12.471-115.063)	0.879±0.26	1.138	0.981	0.949

➢ Solvent extracts concentration from 62.5-500 ppm
➢ LL- Lower limit; UP- Upper limit; Correlation coefficient (R), standard deviation (SD), Chi square test (χ^2).

Table 6 Lethal concentrations (µg/ml) of hexane and dichloromethane extracts of five diffrent seagrasses against *Ae. aegypti* larvae after 48 hours post treatment

Sea grasses	Extracts	LC_{50} LL-UP	LC_{90} LL-UP	Slope±SE	SD	R^2	χ^2
H. uninervis	Hexane	0.474 (0.332-0.678)	1.065 (0.745 -1.522)	3.697±0.07	0.270	0.935	0.097
	DCM	8.251 (1.504-45.275)	705.415 (128.558-387.7)	0.662±0.37	1.510	0.963	0.945
H. ovalis	Hexane	1.342 (0.501- 3.594)	20.259 (7.566-54.243)	1.087±0.21	0.918	0.910	0.672
	DCM	12.488 (2.299-67.824)	226.203 (41.649 -123.567)	1.019±0.37	0.982	0.997	0.995
S. isoetifolium	Hexane	3.587 (0.753-17.100)	34.898 (72.772-165.630)	0.658±0.34	1.520	0.767	0.622
	DCM	34.294 (3.453-340.555)	750.858 (75.612-745.323)	0.956±0.50	1.046	0.999	0.639
C. serrulata	Hexane	0.030 (0.008 -0.117	2.019 (0.520-7.829)	0.702±0.30	1.425	0.955	0.985
	DCM	8.251 (1.504-45.275)	705.415 (128.558-387.699)	0.663±0.37	1.510	0.963	0.945
E. acoroides	Hexane	0.191 (0.066-0.552)	40.518 (5.720-287.015)	0.853±0.23	1.171	0.967	0.953
	DCM	2.433 (0.801-7.389)	53.308 (16.202-175.386)	1.069±0.24	0.936	0.938	0.755

➢ Solvent extracts concentration from 62.5-500 ppm
➢ LL- Lower limit; UP- Upper limit; Correlation coefficient (R), standard deviation (SD), Chi square test (χ^2).

5.2.2. Larvicidal activity of hexane extracts of epiphytic and non-epiphytic portion of *C. serrulata* leaves against third instar larvae of *C. quinquefasciatus* and *Ae. aegypti*:

From the larvicidal activity of hexane extracts of epiphytes and non-epiphytic portion of *C. serrulata* leaves, it was observed that the epiphytic extracts does not exhibit mortality against *Ae. aegypti* and *C. quinquefasciatus* (table. 7, table. 8 graph. 5, graph. 6) whereas, the larvae exhibited mortality in non-epiphytic portion of leaf extracts of *C. serrulata*. The mortality was observed as 88 and 92% at 500 ppm, 52 and 88% at 250 ppm, 48 and 76% at 125 ppm, 40 and 72% at 62.5 ppm against *C. quinquefasciatus*, after 24 hours and 48 hours exposure, respectively (table.7; graph. 5) whereas, the mortality rate was observed as 80 and 88% at 500 ppm, 72% and 80% at 250 ppm, 68 and 76% at 125 ppm, 56% and 72% in 62.5 ppm against *Ae. aegypti* (Table. 8; Graph. 6). Moreover, it was observed that the activity of hexane extracts of non-epiphytic portion of *C. serrulata* leaves extract was found similar to the crude hexane extracts of *C. serrulata* from the above illustrations, it was understood that the larvicidal activity was due to the compounds present in the leaf extracts and not due to epiphytes of *C. serrulata*.

Table 7 Larvicidal activity of hexane extracts of crude, epiphytic and non-epiphytic portion of *C. serrulata* leaves against third instar larvae of *C. quinquefasciatus*

S.No	Hexane extracts of *C. serrulata*	Concentrations (ppm)	Third instar larvae of *C. quinquefasciatus*	
			percentage of mortality after 24 hours exposure(%)	Percentage of mortality after 48 hours exposure(%)
1	Crude leaves	62.5	38	72
		125	44	76
		250	54	86
		500	90	94
2	Non-epiphytic portion of leaves	62.5	40	72
		125	48	76
		250	52	88
		500	88	92
3	Epiphytes	62.5	-	-
		125	-	-
		250	-	-
		500	-	-
4	DMSO	62.5-500 ppm	-	-
5	Untreated water	62.5-500 ppm	-	-

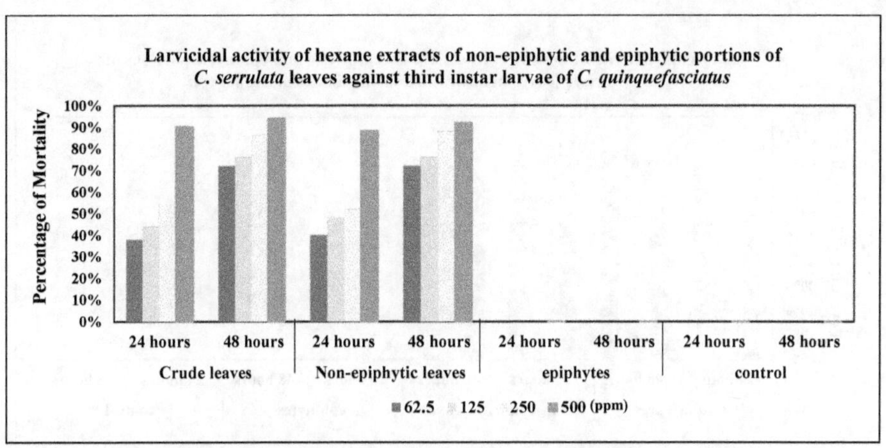

Graph 5 Larvicidal activity of hexane extracts of crude, epiphytic and non epiphytic portions of *C. serrulata* leaves against third instar larvae of *C. quinquefasciatus* after 24 hours and 48 hours exposure

Table 8 Larvicidal activity of crude, epiphytic and non-epiphytic portion of *C. serrulata* leaves against third instar larvae of *Ae. Aegypti*

S.No	Hexane extracts of *C. serrulata*	Concentrations (ppm)	Third instar larvae of *Ae. aegypti*	
			percentage of mortality after 24 hours exposure (%)	Percentage of mortality after 48 hours exposure (%)
1	Crude leaves	62.5	60	72
		125	64	76
		250	76	80
		500	80	86
2	Non- epiphytic portion	62.5	56	72
		125	68	76
		250	72	80
		500	80	88
3	Epiphytes	62.5	-	-
		125	-	-
		250	-	-
		500	-	-
4	DMSO	62.5-500 ppm	-	-
5	Untreated water	62.5-500 ppm	-	-

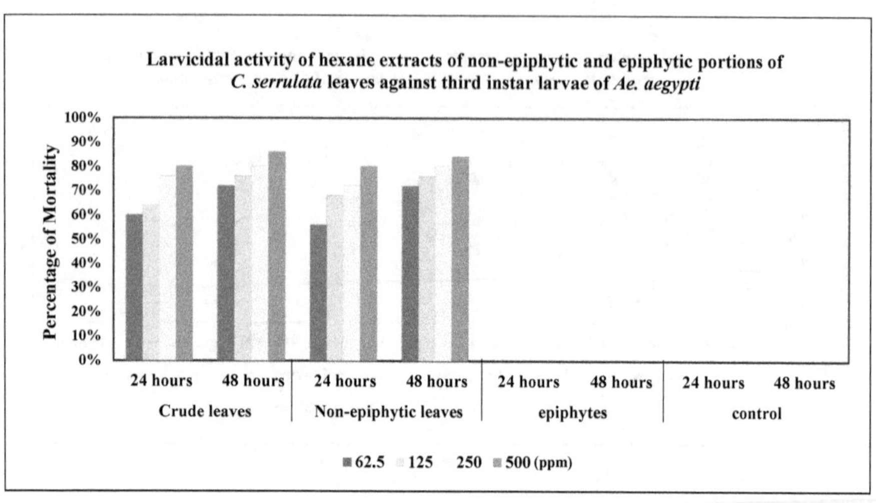

Graph 6 Comparison graph showing larvicidal activity of hexane extracts of crude, epiphytic and non-epiphytic portions of *C. serrulata* leaves against third instar larvae of *Ae. aegypti* after 24 hours and 48 hours exposure

5.3. Isolation and larvicidal activity of fractions of *C. serrulata* leaves:

5.3.1. Isolation and identification of active fraction by bioassay guided fractionation method:

Since the hexane extracts of *C. serrulata* were active against third instar larvae of *C. quinquefasciatus* and *Ae. aegypti*. Therefore, the same extract was subjected to bioassay guided fractionation analysis using column chromatography shown in the Fig. 31. About 1-20 fractions were collected using the solvent system hexane: ethyl acetate in the increasing order of polarity ratio (table. 9). The similar fractions were combined by comparing the 'R$_f$' values and band pattern using TLC over silica gel. As a result, about eight distinctive fractions were obtained from fraction A- H (Fig. 32).

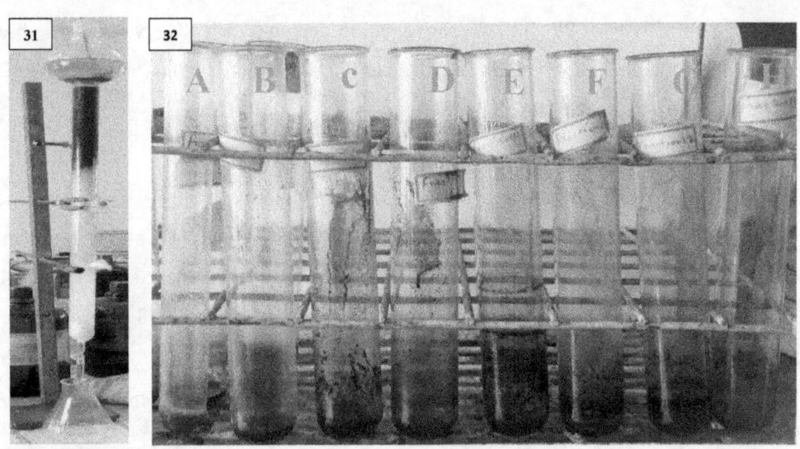

Fig. 31 Bio-guided fractionation assay using column chromatography
Fig. 32 Distinctive fractions from fraction A- fraction H

Table. 9 Solvent system of bio-guided assay fractionation performed in crude hexane extracts of *C. serrulata* leaves

S. No	Solvent System	Fractions Grouped	Combined fractions
1	C_6H_{14} (100%)	1,2,3	Fraction A
2	C_6H_{14}: $CH_3COOC_2H_5$ (9.5:0.5)	4	Fraction B
3	C_6H_{14}: $CH_3COOC_2H_5$ (9:1)	5,6	Fraction C
4	C_6H_{14}: $CH_3COOC_2H_5$ (8:2)	7	Fraction D
5	C_6H_{14}: $CH_3COOC_2H_5$ (3:1)	8,9	Fraction E
6	C_6H_{14}: $CH_3COOC_2H_5$ (2:1)	10,11	Fraction F
7	C_6H_{14}: $CH_3COOC_2H_5$ (1:1)	12,13,14,15	Fraction G
8	$CH_3COOC_2H_5$ (100%)	16,17,18,19,20	Fraction H

Solvents: Hexane- C_6H_{14}; Ethyl acetate- $CH_3COOC_2H_5$

5.3.2. Larvicidal activity of fractions (A-H) of *C. serrulata* leaves:

Larvicidal activity was carried out individually on all the eight fractions (A-H) against third instar larvae of *C. quinquefasciatus* and *Ae. aegypti* (table.10). It shows that, Fraction E was effective against both the larvae whereas other fractions has no effect on larvae. Therefore, Fraction E was separately explained in the table. 11; graph. 7 against *C. quinquefasciatus*. The maximum mortality rate of 80 and 100% was observed at 10 ppm, moderate activity of 42 and 52% found at 5ppm.

The minimum activity of 20 and 24% shown at 2.5 ppm and their LC_{50} values was observed as 0.006 and 0.014 µg/ml; LC_{90} values was found as 0.004 and 0.008 µg/ml (table.12) after 24 hours and 48 hours post treatment, respectively.

Similarly, fraction E found active against *Ae. aegypti* (table.11; graph.7) showed the maximum mortality of 76 and 88% at 10 ppm, moderate activity of 38 and 44% at 5ppm and minimum activity of 16 and 20% at 2.5 ppm and their LC_{50} value was observed as 0.006 and 0.014 µg/ml; LC_{90} value was 0.004 and 0.008 µg/ml (table.13) after 24 hours and 48 hours post treatment, respectively.

Table 10 larvicidal activity of fraction A- fraction E against *C. quinquefasciatus* and *Ae. aegypti* after 24 hours and 48 hours post treatment

Fractions (2.5, 5 and 10 ppm)	Mortality of *C. quinquefasciatus* after 24 and 48 hours	Mortality of *Ae. aegypti* after 24 and 48 hours
Fraction A	-	-
Fraction B	-	-
Fraction C	-	-
Fraction D	-	-
Fraction E	✓	✓
Fraction F	-	-
Fraction G	-	-
Fraction H	-	-
Untreated Water	-	-
DMSO	-	-

✓Mortality; – No mortality

Table 11 larvicidal activity of fraction E against *Ae. aegypti* and *C. quinquefasciatus* after 24 hours and 48 hours post treatment

Sample	Larvae treated	Concentration (ppm)	Percentage of Mortality at 24 hours (%)	Percentage of Mortality at 48 hours (%)
Fraction E	*Ae. aegypti*	2.5	16	20
		5	38	44
		10	76	88
	C. quinquefasciatus	2.5	20	24
		5	42	52
		10	80	100
Untreated Water	*Ae. aegyti and C. quinquefasciatus*	2.5-10	0	0
DMSO	*Ae. aegyti and C. quinquefasciatus*	2.5-10	0	0

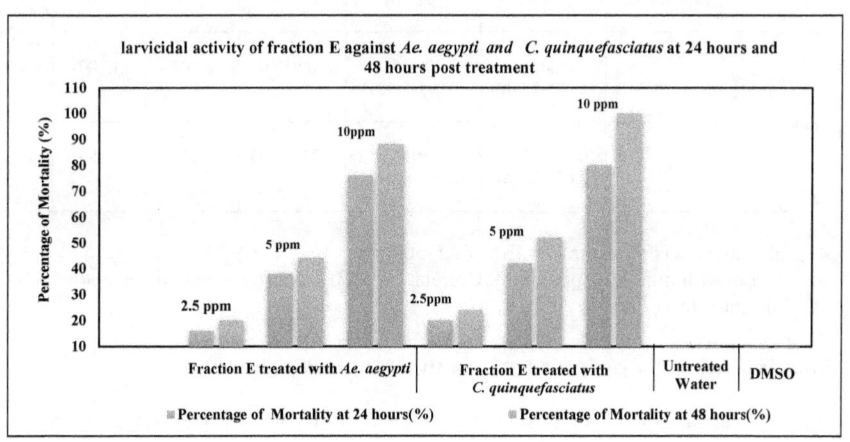

Graph 7 Larvicidal activity of Fraction E against *C. quinquefasciatus* and *Ae. aegypti* after 24 and 48 hours post treatment

Table 12 Susceptibility test of *C. quinquefasciatus* larvae against fraction E of *C. serrulata* after 24 hours and 48 hours post treatment

Fraction	Hours of exposure (h)	LC$_{50}$ LL-UP (µg/ml)	LC$_{90}$ LL-UP (µg/ml)	Slope±SE	SD	R^2	χ^2
Fraction E	24	0.006 (0.004-0.008)	0.014 (0.010-0.020)	3.233±0.07	0.309	0.997	0.847
	48	0.004 (0.003-0.006)	0.008 (0.006-0.011)	4.754±0.05	0.210	0.983	0.531

Table 13: Susceptibility test of *Ae. aegypti* larvae against fraction E of *C. serrulata* after 24 hours and 48 hours post treatment

Fraction	Hours of exposure(h)	LC$_{50}$ LL-UP (µg/ml)	LC$_{90}$ LL-UP (µg/ml)	Slope±SE	SD	R^2	χ^2
Fraction E	24	0.006 (0.004-0.009)	0.015 (0.011-0.021)	3.259±0.07	0.307	1.000	0.929
	48	0.004 (0.003-0.006)	0.008 (0.006-0.011)	4.735±0.05	0.210	0.983	0.531

- Solvent extracts concentration from 62.5-500 ppm
- LL- Lower limit; UP- Upper limit; Correlation coefficient (R), standard deviation (SD), Chi square test (χ^2).

5.3.3. Isolation of the compound 'A' and 'B' of *C. serrulata* leaves:

The two compounds were observed in the Fraction E of TLC analysis (Fig.33), therefore, both the compounds were eluted separately using sub column bio guided fractionation using the hexane: ethyl acetate solvents (Fig. 34). About 1 to14 subfractions were collected, among that two single band was found in sub-fractions 10,11 with the same 'R$_f$' value 0.4 (fig. 35; table.14). So, they were combined, washed and crystallised using acetone, showing colourless powder (mp 145° C) and considered as compound 'A' whereas, another single band was found in sub-fractions 13,14 with the 'R$_f$' value 0.33 (fig. 35; table. 14), combined, crystallised and considered as compound 'B'.

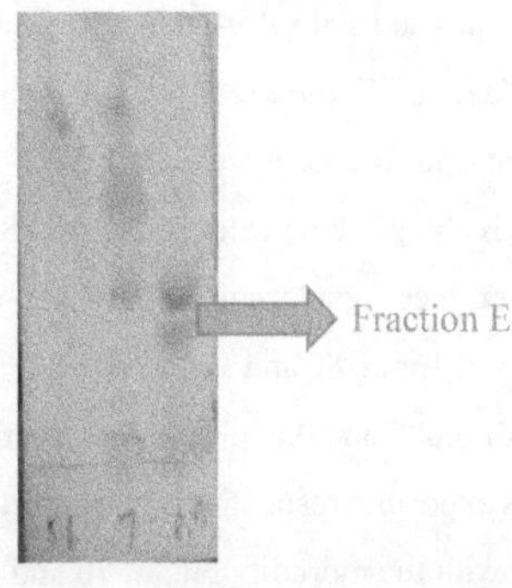

Fig 33 TLC pattern of fraction E of *C. serrulata* leaves

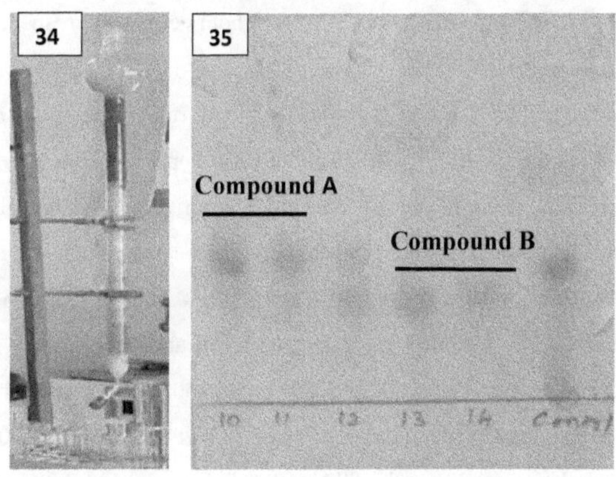

Fig 34 Photograph showing sub column fractionation of fraction E
Fig 35 TLC plate showing isolation of single compounds 'A' and 'B'

Table 14 Bio-guided assay of single isolated compounds

Sub-fractions	Single isolated Compound	Rf value
10,11	Compound A	0.46
13,14	Compound B	0.33

5.3.4. Larvicidal activity of the compound 'A' and compound 'B' of *C. serrulata* leaves:

Among the two compounds 'A' and 'B', Compound 'A' was found to be more efficient against third instar larvae of *C. quinquefasciatus* and *Ae. aegypti*. whereas, no mortality was found in compound 'B' against both the larvae (table. 15; table. 16). Results from the table. 15; graph. 8; table. 17 showed the mortality rate of *C. quinquefasciatus* 20 and 24% at 1 ppm, 42 and 52% at 2.5 ppm, 80 and 100% at 5ppm with the LC_{50} values of 0.004 and 0.006 µg/ml; LC_{90} values of 0.011 and 0.019 µg/ml, respectively, $\chi2$ value calculated as 0.585 and 0.871 after 24 and 48 hours exposure, respectively. However, these results can be very well compared to the azadirachtin with the mortality rate of 54 and 70% at 1ppm, 80 and 100 % at 2.5-5 ppm with the LC_{50} values of 0.002 and 0.003 µg/ml; LC_{90} values of 0.007 and 0.016 µg/ml, respectively, $\chi2$ value calculated as 0.962 and 0.927 after 24 and 48 hours exposure, respectively. The mortality rate was even compared with chemical compound temephos with the mortality rate of 70 and 84% at 1ppm, 100% at 2.5-5 ppm with the LC_{50} values of 0.001 and 0.002 µg/ml; LC_{90} values of 0.004 and 0.006 µg/ml, respectively, $\chi2$ value was observed as 0.962 and 0.927 after 24 and 48 hours exposure, respectively.

Similarly, results from the table. 16; graph. 9; table. 18, the mortality rate of *Ae. aegypti* found as 20 and 24% at 1 ppm, 42 and 52% at 2.5 ppm, 70 and 90% at 5ppm with the LC_{50} values of 0.040 and 0.050 µg/ml; LC_{90} values of 0.080 and 0.140 µg/ml, respectively, $\chi2$ value calculated as 0.531 and 0.847 after 24 and 48 hours exposure, respectively. These results can be compared to the azadirachtin with the mortality rate of 46 and 70% at 1ppm, 70 and 90 % at 2.5-5 ppm with the LC_{50} values of 0.020 and 0.030 µg/ml; LC_{90} values of 0.060 and 0.070 µg/ml, respectively, $\chi2$ value calculated as 0.876 and 0.961 after 24 and 48 hours exposure, respectively. The mortality rate was even compared to temephos the mortality rate of 70 and 86% at 1ppm, 100% at 2.5-5ppm with the with the LC_{50} values of 0.001 and 0.002 µg/ml; LC_{90} values of 0.004 and 0.006 µg/ml, respectively, $\chi2$ value was observed as 0.571 and 0.863 after 24 and 48 hours exposure, respectively.

Further, the purity of the compound 'A' will be further confirmed by GC-MS analysis, if the compound is of low polarity, the identification of the compound / compounds will be confirmed by spectroscopic methods UV-Vis spectroscopy, FTIR, ^{1}HNMR *etc.*

Table 15 Larvicidal activity of compoundS 'A' and 'B' against third instar larvae of *C. quinquefasciatus* after 24 hours and 48 hours of treatment

S.No	Compounds	Mortality at 24 hours (%)	Mortality at 48 hours (%)
1	Compound A (2.5-10ppm)	20% (1 ppm) 42% (2.5 ppm) 80% (5 ppm)	24% (1 ppm) 52% (2.5 ppm) 100% (5 ppm)
2	Compound B (2.5-10ppm)	0	0
3	Azadiractin	54% (1 ppm) 80% (2.5 ppm) 100% (5 ppm)	70% (1 ppm) 100% (2.5 ppm) 100% (5 ppm)
4	Temephos (2.5-10ppm)	70% (1ppm) 100% (2.5-5ppm)	84% (1 ppm) 100% (2.5-5ppm)

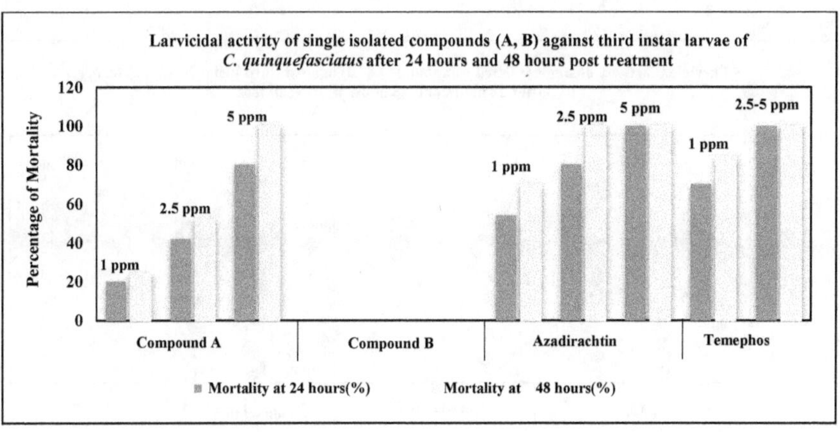

Graph 8 Larvicidal activity of the compounds 'A' and compound 'B' against third instar larvae of *C. quinquefasciatus* after 24 and 48 hours post treatment

Table 16 Larvicidal activity of single isolated Compound 'A' and compound 'B' against third instar larvae against *Ae. aegypti* after 24 hours and 48 hours post treatment

S.No	compounds	Percentage of mortality at 24 hours (%)	Percentage of mortality at 48 hours (%)
1	Compound A (2.5-10ppm)	20% (1 ppm) 42% (2.5 ppm) 70% (5 ppm)	24%(1 ppm) 52% (2.5 ppm) 90% (5 ppm)
2	Compound B (2.5-10ppm)	0	0
3	Azadirachtin	46% (1 ppm) 70% (2.5 ppm) 90% (5 ppm)	70% (1 ppm) 90% (2.5 ppm) 100% (5 ppm)
4	Temephos (2.5-10ppm)	70% (2.5 ppm) 100% (2.5-5 ppm)	86% (2.5 ppm) 100% (2.5-5 ppm)

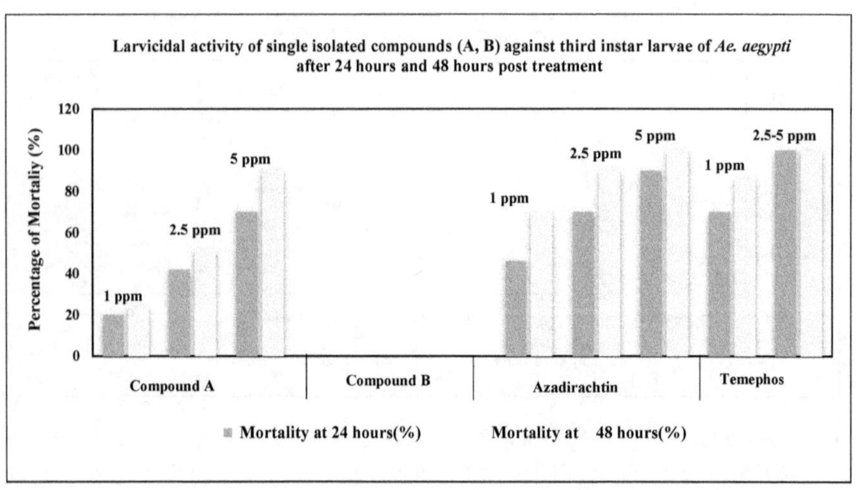

Graph 9 Larvicidal activity of single isolated compound 'A' and compound 'B' against third instar larvae of *Ae. aegypti* after 24 and 48 hours post treatment

Table 17 Lethal concentrations (µg/ml) of compound 'A', Azadirachtin and temephos treated against third instar larvae of *C. quinquefasciatus* after 24 hours and 48 hours post Treatment

Sea grasses	Hours of exposure(h)	LC_{50} LL-UP	LC_{90} LL-UP	Slope±SE	SD	R^2	χ^2
Compound A	24	0.004 (0.003-0.006)	0.011 (0.007-0.015)	3.289±0.07	0.304	0.972	0.585
	48	0.006 (0.004- 0.009)	0.019 (0.013-0.029)	2.600±0.09	0.384	0.997	0.871
Azadirachtin	24	0.002 (0.002-0.004)	0.007 (0.004-0.011)	2.873±0.09	0.348	0.999	0.960
	48	0.003 (0.002-0.006)	0.016 (0.009-0.031)	1.759±0.13	0.569	0.929	0.927
Temephos	24	0.001 (0.001-0.002)	0.004 (0.002-0.011)	1.638±0.20	0.371	0.671	0.571
	48	0.002 (0.001-0.003)	0.006 (0.004-0.010)	2.698±0.11	0.610	0.671	0.863

➢ Solvent extracts concentration from 62.5-500 ppm
➢ LL- Lower limit; UP- Upper limit; Correlation coefficient (R), standard deviation (SD), Chi square test (χ^2).

Table 18 Lethal concentrations (µg/ml) of compound 'A', Azadirachtin and temephos treated against third instar larvae of *Ae. aegypti* at 24 hours and 48 hours post Treatment

Sea grasses	Hours of exposure(h)	LC$_{50}$ LL-UP	LC$_{90}$ LL-UP	Slope±SE	SD	R^2	χ^2
Compound A	24	0.040 (0.030-0.060)	0.080 (0.060-0.110)	4.754±0.05	0.210	0.983	0.531
	48	0.050 (0.040-0.080)	0.140 (0.100-0.200)	3.233±0.07	0.309	0.997	0.847
Azadirachtin	24	0.020 (0.010-0.030)	0.060 (0.030-0.110)	2.300±0.13	0.435	0.952	0.876
	48	0.030 (0.020-0.040)	0.070 (0.050-0.100)	3.148±0.08	0.318	0.999	0.961
Temephos	24	0.010 (0.00-0.020)	0.030 (0.010-0.130)	1.208±0.293	0.828	0.671	0.728
	48	0.010 (0.010-0.030)	0.050 (0.030-0.100)	2.210±0.14	0.452	0.671	0.927

➢ Solvent extracts concentration from 62.5-500 ppm
➢ LL- Lower limit; UP- Upper limit; Correlation coefficient (R), standard deviation (SD), Chi square test (χ^2).

5.4. Characterisation and identification of compound 'A' of *Cymodoceae serrulata* leaves:

5.4.1. UV-Visible Spectroscopy:

The intensity of the UV-Vis spectrum of compound 'A' was found maximum at 255 nm (Fig. 36). From the previous literature, Govindarajan and Sarada, 2011 the absorption spectra was found at 257 nm shows the presence of mixture of phytosterols. Therefore, in the present study also it could be predicted as phytosterol.

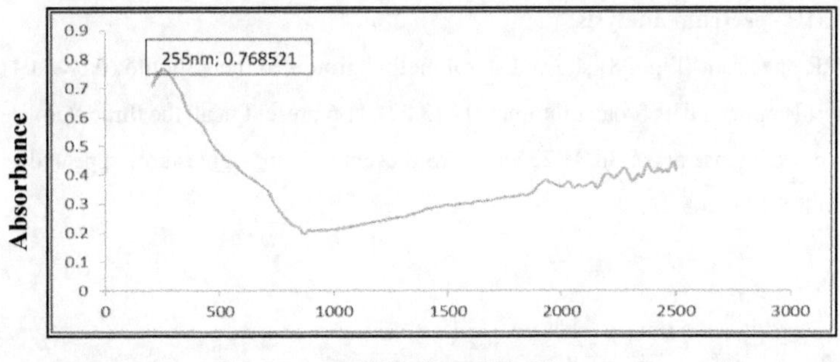

Wavelength (nm)

Fig 36 UV-Vis spectrum of isolated compound 'A' of *C. serrulata* leaves

5.4.2. FTIR analysis:

The FTIR spectrum of compound 'A' (Fig. 37) showed peaks at 3430cm^{-1} (O-H stretch) indicates presence of hydroxyl groups, 2938 and 2862 cm^{-1} (C−H stretch) indicates the presence of alkane group, 1640 cm^{-1} (C=C stretch) 1640 cm^{-1} indicates diketones, 1468 cm^{-1} (C-H bending) shows the presence of aromatic ring, 1375 cm^{-1} (CH$_3$ bending) indicates nitro groups, 1161, 1101 and 1053 cm^{-1} (C−O stretch) indicates presence of alkyl amine 966 and 840 cm^{-1} (=C−H bending) indicates carboxylate groups and aromatic compounds (suttiarporn *et al.*, 2015; Deepasree *et al.*, 2012)

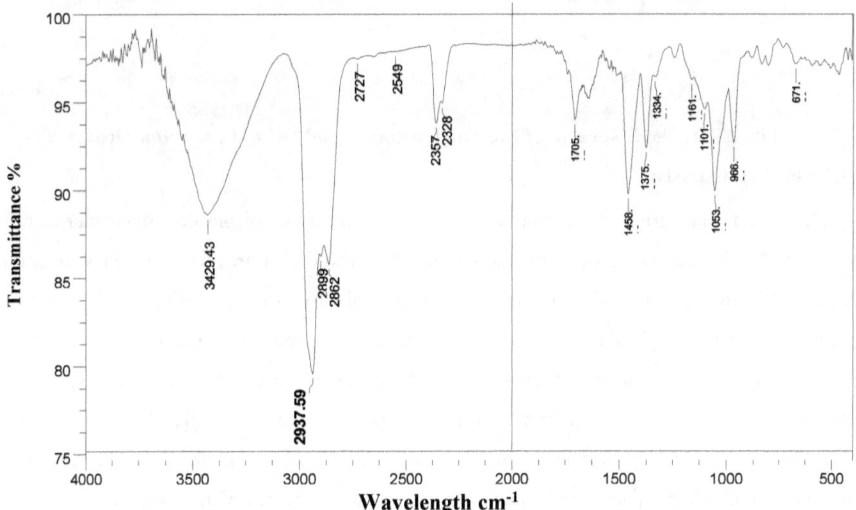

Wavelength cm^{-1}

Fig. 37 FTIR spectrum of isolated compound 'A' of *C. serrulata* leaves

5.4.3. ^{1}H NMR spectrum analysis:

The ^{1}H NMR spectrum (Fig. 38) showed sterol methyl groups in the region δ (0.68-1.04). H-3 of the three sterols appeared as broad multiplet at δ (3.52). H-6 present in all the three sterols appeared as a triplet, more intense at δ 5.36. H-22 and H-23 present only in stigmasterol appeared as smaller multiplets at δ 5.15 and 5.05.

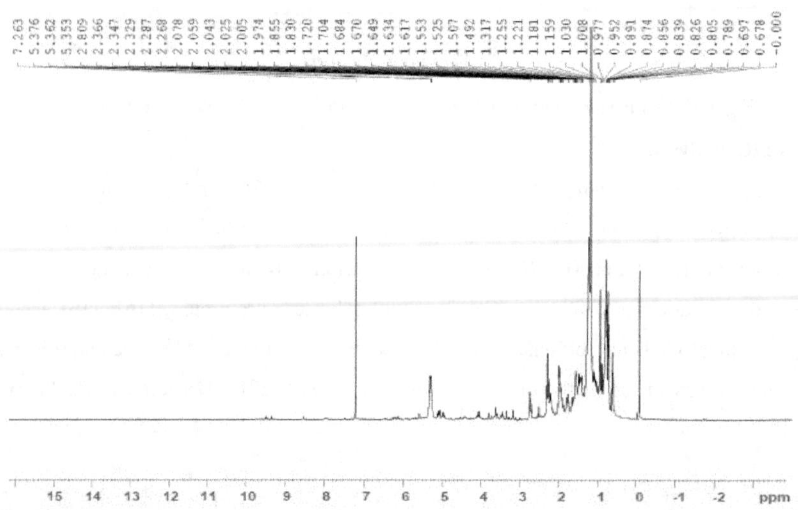

Fig 38 ^{1}H NMR spectra of the isolated compound 'A' of *C. serrulata* leaves

5.4.4. GC-MS analysis:

The GC-MS chromatogram of compound 'A' of *C. serrulata* leaves displayed 3 major sterols from Fig. 39; table.19. The mass spectra of stigmasterol (1) showed 27.1 area percentage and retention time at 29.05 min, M$^+$ at m/z 412 corresponding to molecular formula $C_{29}H_{48}O$ (Fig. 40). The mass spectra of campesterol (2) showed 27.44 area percentage and retention time at 29.37 min, M$^+$ at m/z 400 corresponding to the molecular formula $C_{28}H_{48}O$ (Fig. 41). The mass spectra of betasitosterol (3) showed M$^+$ at m/z 414 corresponding to the molecular formula $C_{29}H_{50}O$ (Fig. 42). The spectroscopic data given above was comparable with the literature reports by suttiarporn *et al.*, 2015). There were some minor compounds reported with its area percentage and retention time from fig. 39; table. 19 decane 1.18 % at 8.31 min, pentadecane 2,6,10,14-tetramethyl

1.16% at 11.16 min, 2-pentadecaone, 6,10,14-trimethyl 1.17 % at 16.35 min, retinol 1.49 % at 27.72 min, desmosterol 1.81% at 28.20 min, 26-nor -5 cholosten-5-cholosten 3Beta-ol-25one 6.18% at 28.51 min.

However, many previous kinds of literature declared that the presence of three sterols stigmasterol, campesterol and betasitosterol together occurred in many plants (yucei *et al.*, 2017; suttiarporn *et al.*, 2015; sogan *et al.*, 2018; wang *et al.*, 2011; leone *et al.*, 2016; lorensi *et al.*, 2019; Sorenson and Sullian, 2007). Moreover, it was also reported that these group of sterols are difficult to be separated through ordinary column chromatography or TLC but can be analysed through HPLC, LCMS or by GCMS methods

Rahman *et al.*, 2008 isolated betasitosterol from petroleum ether leaf extracts of *Abutilon indicum* has proven to have effective larvicidal activity against the early forth instar larvae of *Ae. aegypti*, *An. stephensi* and *C. quinquefasciatus* with LC_{50} values of 11.49, 3.58 and 26.67 ppm at 24 hours respectively. Gade *et al.*, 2017 isolated stigmasterol and n-hexacosonal from *Chromolaena odorata* found effective against the third instar larvae of *C. quinquefasciatus* and *Ae. aegypti* but, stigmasterol in particular found effective against *C. quinquefasciatus* and *Ae. aegypti* with the LC_{50} value of 70.84 mg/ml and 25 mg/ml after 24 hours.

However, there is no particular reports for larvicidal activity of campesterol from plants but lorensi *et al.*, 2019 has reported in earlier studies, the activities of betasitosterol, campesterol and stigmasterol were isolated from *Prasiola crispa* and treated against *Nauphoeta cinerea* coackroaches. The mechanism of action is through acetylcholinesterase inhibition leading to hyper excitability of the olfactory system. They have found that the compounds induced complete inhibition of the cockroach leg muscles titch by inhibiting the acetylcholinesterase activity by blocking the neuromuscular junction transmission. The activity was found in the order betasitosterol > campesterol > stigmasterol. This experiment showed the possibility of campesterol also has appreciable mosquito larvicidal activity.

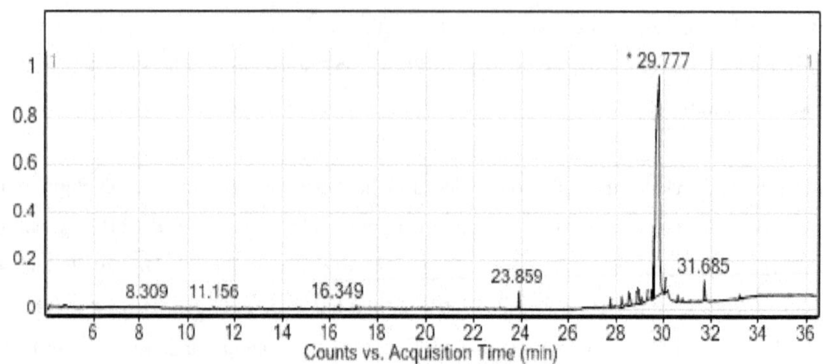

Fig.39 GCMS analysis of compound 'A' of *C. serrulata* leaves

Table.19: GC-MS analysis of compound 'A' of *C. serrulata* leaves

RT (Minutes)	Compound Name	Molecular Formula	Molecular Mass (Dalton)	Area%
8.31	Decane, 5-ethyl-5-methyl-	$C_{13}H_{28}$	184.2	1.18
11.16	Pentadecane, 2,6,10,14-tetramethyl-	$C_{19}H_{40}$	268.3	1.16
16.35	2-pentadecanone,6,10,14-trimethyl-	$C_{18}H_{36}O$	268.3	1.17
27.72	Retinol, acetate	$C_{22}H_{32}O_2$	328.2	1.49
28.20	Desmosterol	$C_{27}H_{44}O$	384.3	1.81
28.51	26-Nor-5-Cholesten-3.beta-ol-25-one	$C_{26}H_{42}O_6$	386.3	6.18
29.05	Stigmasterol	$C_{29}H_{48}O$	412.4	27.10
29.37	Campesterol	$C_{28}H_{48}O$	400.4	27.44
30.09	Beta.-Sitosterol	$C_{29}H_{50}O$	414.4	28.54

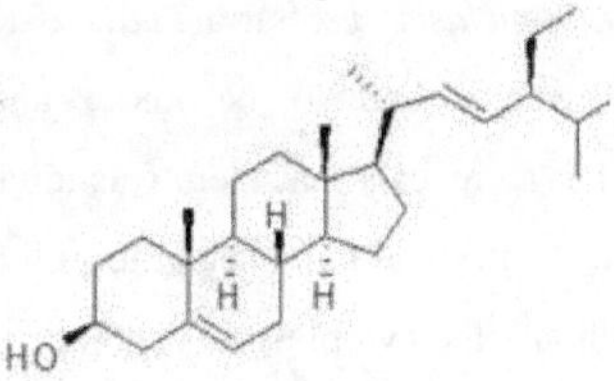

Fig. 40. Stigmasterol (1)

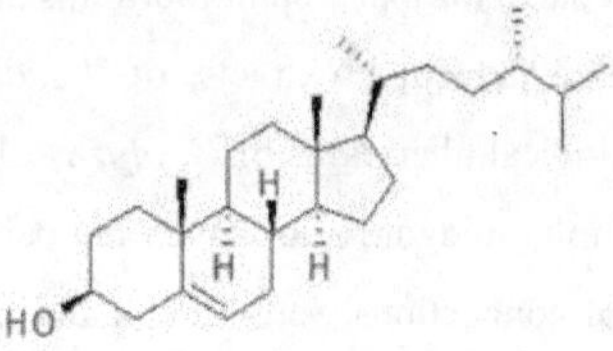

Fig. 41. Campesterol (2)

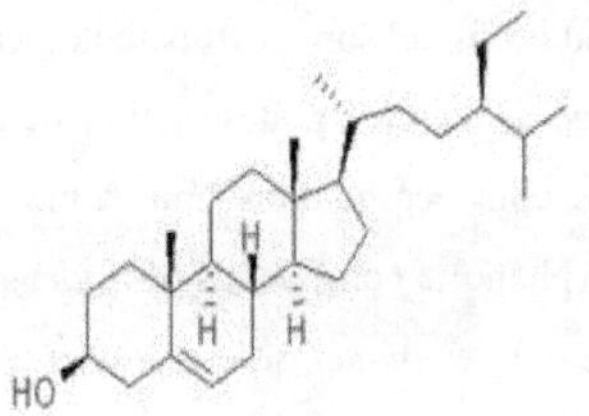

Fig. 42. Betasitosterol (3)

5.5. Histopathological alterations of third instar larvae of *C. quinquefasciatus* against isolated compound 'A' of *C. serrulata* leaves

Third instar larvae of *C. quinquefasciatus* were exposed to untreated water, temephos and Compound 'A' of *C. serrulata* leaves were analysed histopathological studies (Fig. 43a; Fig. 43b and Fig. 44c). From the Fig. 44c compound 'A' showed morphological changes when compared to temephos Fig. 43b. It was observed that disruption of the circular muscle layers (EP) and adipose cells. The midgut epithelial columnar cells (CC), the Nucleus (N) were found to be enlarged and vacuolated, peritrophic membrane (pM) and food bolus (FB) was affected whereas, muscle layers, midgut region and food bolus enclosed in a thick peritrophic membrane were found normal in the untreated larvae.

Similar research was previously reported by Farida *et al.*, 2017 entomopathogenic fungus *Beauveria bassiana* treated against fourth instar larvae of *C. pipiens* and the histological studies showed malformation and alterations in the body parts such as cuticle was disturbed and disruption of midgut regions. Ragavendran *et al.*, 2019 studied that the mycelia of *Penicillium* Sp. disintegrated the abdomen region specifically midgut and caecae, epithelial region was disorganised and disappeared in fourth instar larvae of *C. quinquefasciatus*. Sarral *et al.*, 2017 was reported that aqueous extracts of *Eucalyptus globus* were found to be toxic, and the histopathological alterations of *C. pipens* of fourth instar larvae reveals that there was a damage occurred in the midgut regions with the periodical hours like intestinal cells began to pull away, intercellular connections were broken and finally after 24 hours the cytoplasmic contents moved on to the intestinal lumen and completely leads to death.

Al- Mekhlafi, 2018 reported that crude extracts of *Carum copticum* seeds was effective against fourth instar larvae of *C. pipiens* and observed some cytopathological changes such as destruction of midgut epithelial cells, existence of vescicles, destruction of microvilli, swollen cells. cell disintegration, chromatin and nucleoli degradation and some internal contents were lost. Jiraungkoorskul, 2016 reported the phenolic compounds of *Andrographis paniculata* leaf of was effective against fourth instar larvae of *C. quinquefasciatus*, under histopathological studies, it was observed that base membrane was separated from the epithelial cells, brush border disruption, several vesicles and certain cytoplasmic masses were observed.

Al-Mehmadi and Al-Khalaf, 2010 reported *Melia azedarach* plant extracts was effective against the larvae of *C. quinquefascitus*, which has affected the columnar epithelial cells of midgut and it was detached from the peritrophic membrane. Moreover, Devan *et al.*, 2015, reported that natural compound called catechin, a phenolic derivative has affected the epithelial cells, disrupted the columnar cells, some of the cells were found to be dislodged and bulging of the middle portion of the gut.

Aminah *et al.*, 2001 reported previously lipid compound saponins become corrosive in the digestive tract due to the reduction of surface tension. Like that, in the present study also, lipid compound phytosterols of *C. serrulata* leaf found toxic against *C. quinquefasciatus* larvae and the histopathological studies displayed the damage in muscle layers, the epithelial membrane, other cell components of midgut cells and finally affected the food bolus. They are compared with synthetic compound temephos for the experimental evidence. However, the untreated larvae exhibited normal in the muscle layers, midgut epithelial cells, cell membranes and cytoplasmic region.

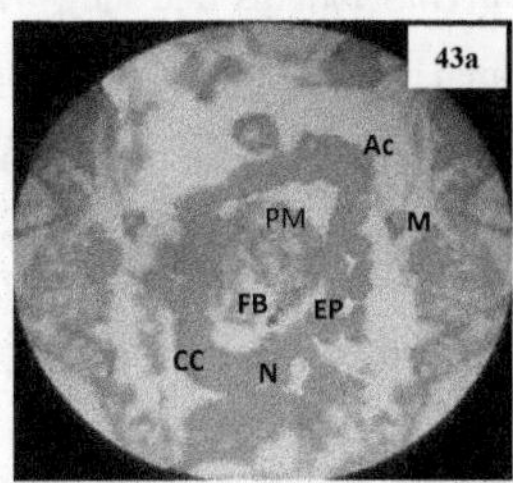
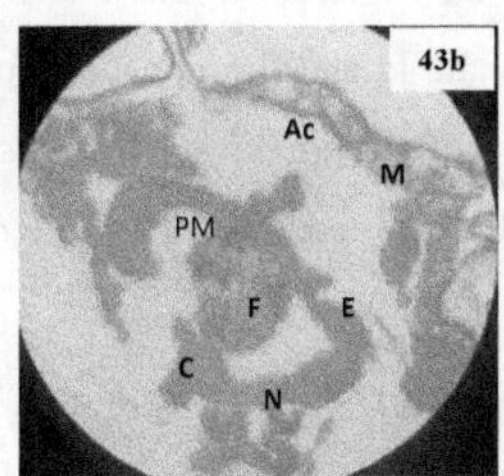
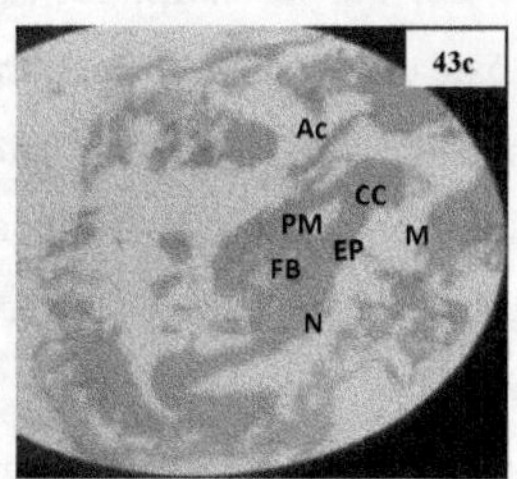

Fig 43 Cross section via midgut part of the third instar larvae of *C. quinquefasciatus* against untreated water (43a); temephos (43b); treated with compound 'A'(43c)

5.6. Effect of Phytosterol mixtures (compound 'A') against non-target organisms of *Poecilia reticulata* and *Gambusia affinis*:

The toxic level of Phytosterol mixtures of *C. serrulata* leaf extracts were tested against non-target organisms on predatory fishes *Gabussia affinis* and *Poecilia reticulata* (Fig. 44; Fig. 45). The phytosterol mixtures treated against *G. affinis* and *P. reticulata* with the LC_{50} value was found as 0.032 and 0.094 µg/ml and LC_{90} value was found as 0.102 and 0.557 µg/ml was summarised in table. 20. They were compared with the azadirachtin with the LC_{50} and LC_{90} value of 0.003 and 0.054 µg/ml and temephos with the LC_{50} and LC_{90} value of 0.001 and 0.013 µg/ml, after 24 hours, respectively.

The suitability index/ predatory safety factor (SI/PSF) of a phytosterols (Compound A) was summarised in (table. 21), The SI/PSF of *G. affinis* and *P. reticulata* when exposed to compound 'A' was found as 4 and 9.27, when they exposed to Azadiractin it was found as 1.5 and 7.71 and when exposed to temephos it was 1 and 3.25. It was also observed that there was no harm to the predatory fishes when exposed to predatory fishes till 5ppm and it is safe to use in the environment.

Similar research was previously studied by Sivagnaname and kalyanasundaram, 2004 reported methanolic extracts of *Atlantia monophylla* (Rutaceae) were safe on predatory fishes *Gambusia affinis*, *Poecilia reticulata* and aquatic bug *Diplonychus indicus* and found non-toxic upto 5 mg/ml. Reegan *et al.*, 2014 isolated a compound Nilotin from *Limonia acidissima*

(Rutaceae) was recommended safe on *Gabussia affinis* and *Diplonychus indicus* and they were non-toxic upto 2 ppm. Maheswaran and Iganachimuthu, 2012 prepared a novel drug Poneem, is a combination of *Azadirachta indica* and *Pongamia glabra* was effective against *Ae. aegypti* and *Ae. albopictus* larvae and the effect on predatory fish *Gabussia affinis* and predatory bug *Diplonychus indicus* were reported harmless.

Pavela and Govindarajan, 2017 isolated two compounds Germacrene –D-4 and α- cadinol from leaves of *Zanthoxylum monophyllum* found active against *An. stephensi, Ae. aegypti and C. tritaeniorhynchus* and the effect on non-target organism *Gambusia affinis* was least sensitive, SI and PSF was suitable. Govindarajan and Benelli *et al.*, 2016 reported that the essential oil of *Artemissia abisinthium* was found presence of three compounds (E)-β-farnesene, (Z)-en-yn-dicycloether and (Z)-β-ocimene were tested against six mosquito vectors *An. Stephensi, An. subpictus, Ae. aegypti, Ae. albopictus, C. quinquefasciatus* and *C. tritaeniorhynchus*. It was observed that the three compounds were toxic to all the six vectors and the effect of non-target organism such as *Chironomous circumdatus, Anisops bouvieri* and *Gambusia affinis* was moderately less toxic.

However, in the present study the phytosterol mixtures was found non effective against predatory fishes upto 5 ppm and the suitability index factor was also favourable.

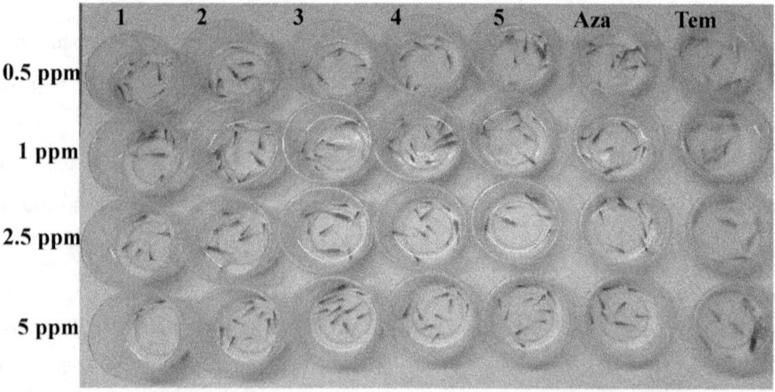

Fig 44 Non-target effect of phytosterol mixtures of *C. serrulata* leaf extracts against *Gambusia affinis* after 24 hours post treatment

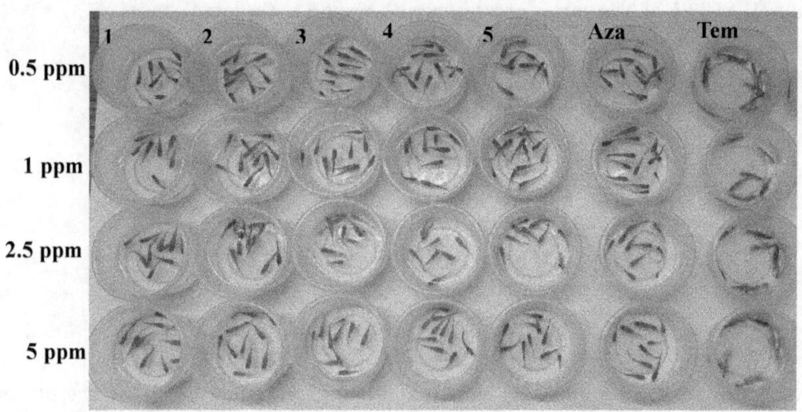

Fig 45 Non-target effect of phytosterol mixtures of C. *serrulata* leaf extracts against *Poecilia reticulata* after 24 hours post treatment

Table 20 Effect of phytosterol mixtures against non-target organism *Gambusia affinis* and *Poecilia reticulata*

Treatment	Gambusia affinis		Poecilia reticulata	
	LC_{50}	LC_{90}	LC_{50}	LC_{90}
Phytosterol mixtures	0.032	0.102	0.094	0.557
Azadirachtin	0.003	0.054	0.003	0.054
Temephos	0.001	0.013	0.001	0.013

Table 21 SI/PSF of Non-target organisms with respect to the larval stages of
C. quinquefasciatus exposed to Beta sitosterol and Campesterol

Treatment	Predatory Species	
	Gambusia affinis	_Poecilia reticulate_
Phytosterol mixtures	4	9.27
Azadirachtin	1.5	7.71
Temephos	1	3.25

5.7. _In silico_ molecular docking analysis of _ace1_ receptor of _Ae. aegypti_ and _C. quinquefasciatus_ against Phytosterol mixtures:

The binding interaction and binding energy of homology model of _ace1_ receptor of _C. quinquefasciatus_ with ligands were viewed and tabulated in Table. 22; Fig. 46A; Fig. 46B; Fig. 46C; Fig. 46D; Fig. 46E. The _ace1_ receptor interacted with stigmasterol was found as -555.33 and -7.21 Kcal/Mol, interaction with campesterol was -560.05 and -7.29 Kcal/Mol, interaction with betasitosterol as -551.33 and -7.63 Kcal/Mol. These interactions were compared with azadirachtin -120.33 and -7.14 Kcal/Mol and temephos as -676.43 and -7.72 Kcal/Mol.

Similarly, The binding energy and binding interaction of homology model of _ace1_ receptor of _Ae. aegypti_ with ligands were viewed and summarised in Table. 22; Fig. 47A; Fig. 47B; Fig. 47C; Fig. 47D; Fig. 47E. The _ace1_ receptor interacted with stigmasterol was found as -7.20 and - 439.67 K cal/Mol, interaction with campesterol found as -452.98 and -7.39 Kcal/Mol, interaction with betasitosterol was found as -457.40 and -8.04 Kcal/Mol, interaction with azadiractin was found as -121.33 and -7.13 Kcal/Mol and interaction with temephos was found as -577.66 and -8.05 Kcal /Mol.

In the previous study, Similar study was reported by Reegan _et al._, 2016, isolated niloctin compound docked with the _ace1_ receptor of _Ae. aegypti_. The binding energy and inhibition constant value was found as -8.4Kcal/Mol and 696.118 µM. This was found more efficient when compared to synthetic insecticides temephos as -4.75 Kcal/Mol.

Table 22 *In silico* molecular Docking analysis of Phytosterol mixtures (Stigmasterol, Beta sitosterol, Campesterol) and control (azadirachtin and temephos) against homology model of *ace1* receptor of *Ae. aegypti* and *C. quinquefasciatus*

Homology model of *ace1* receptor	Ligand	Cluster analysis	Fullness Kcal/Mol	Estimated Δ G Kcal/Mol
ace1 Model Id_5ydi.2.A (*Ae. aegypti*)	Stigmasterol	18	-439.67	-7.20
	Campesterol	7	-452.98	-7.39
	Betasitosterol	1	-457.40	-8.04
	Azadirachtin	4	-121.33	-7.13
	Temephos	20	-577.66	-8.05
ace1 Model Id_5x61.1 (*C. quiquefascitus*)	Stigmasterol	2	-555.33	-7.21
	Campesterol	0	-560.05	-7.29
	Betasitosterol	3	-551.33	-7.63
	Azadirachtin	5	-120.33	-7.14
	Temephos	17	-676.43	-7.72

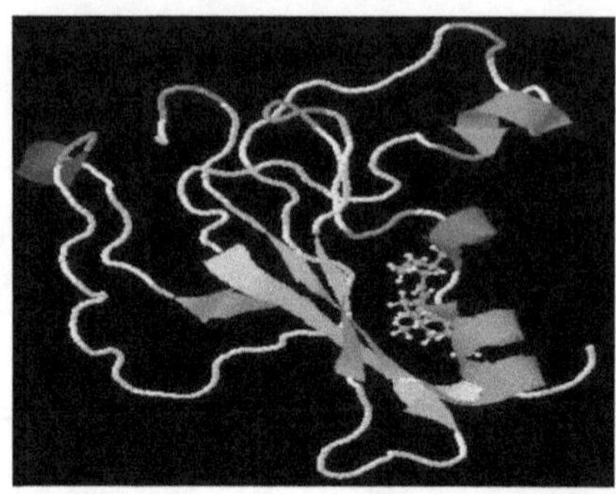

Fig. 46A. Interaction with stigmasterol

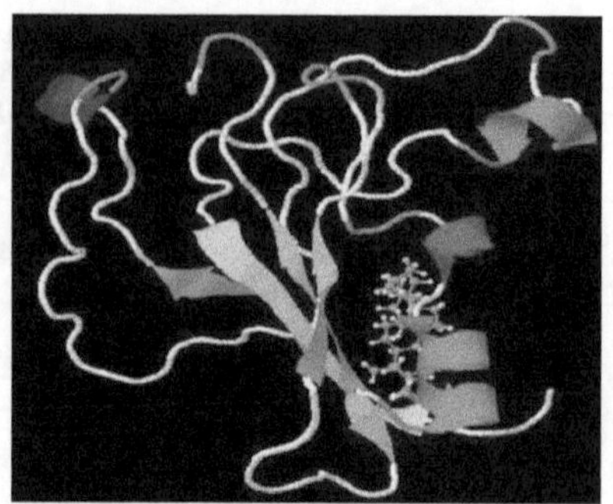

Fig. 46B. Interaction with campesterol

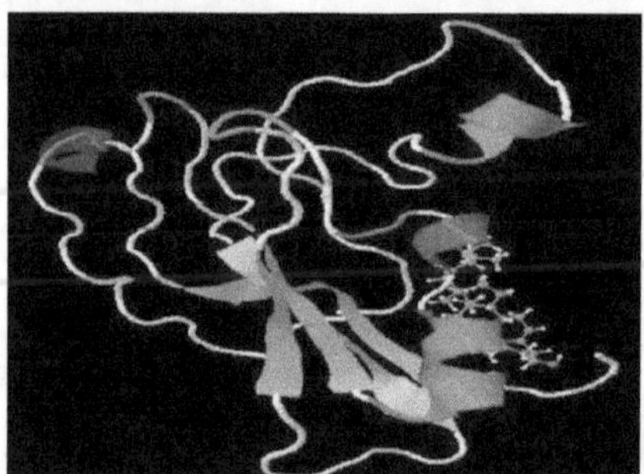

Fig. 46C. Interaction with betasitosterol

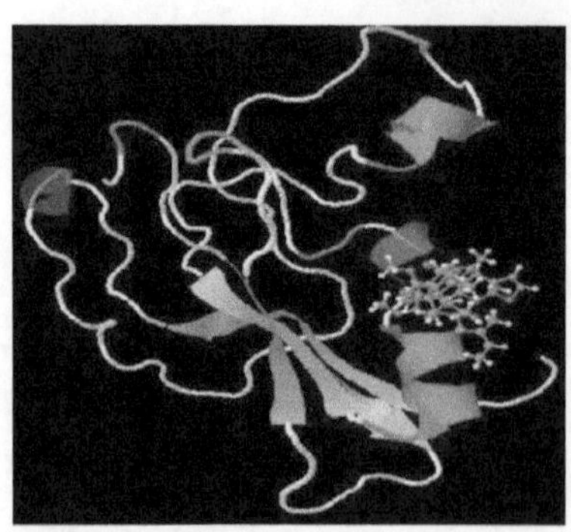

Fig. 46D. Interaction with azadirachtin

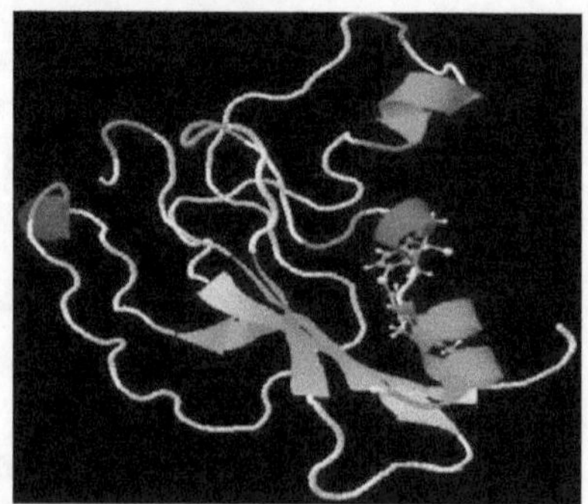

Fig. 46E. Interaction with temephos

Fig. 46 *In silico* molecular Docking analysis of homology model of *ace1* receptor of *C. quinquefasciatus* interacted with Stigmasterol (46A); Campesterol (46B); Beta sitosterol (46C); Azadirachtin (46D); Temephos (46E)

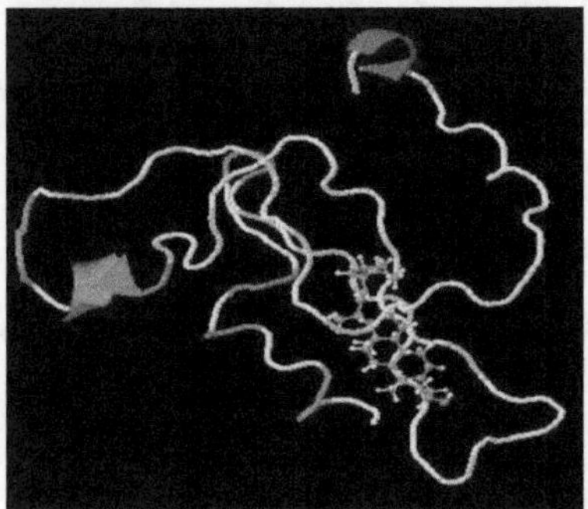

Fig. 47A Interaction with stigmasterol

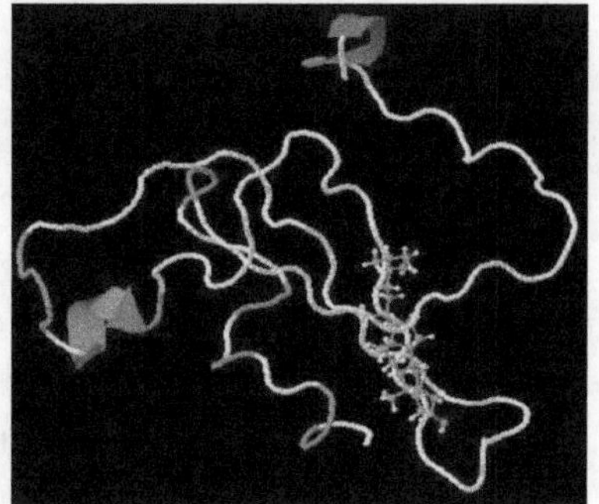

Fig. 47B Interaction with campesterol

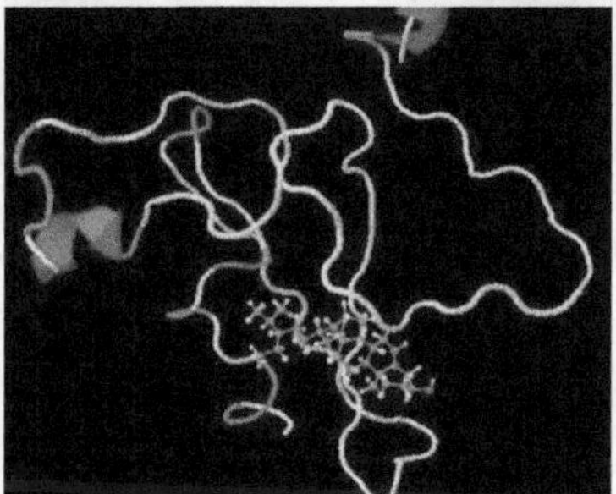

Fig. 47C Interaction with Betasitosterol

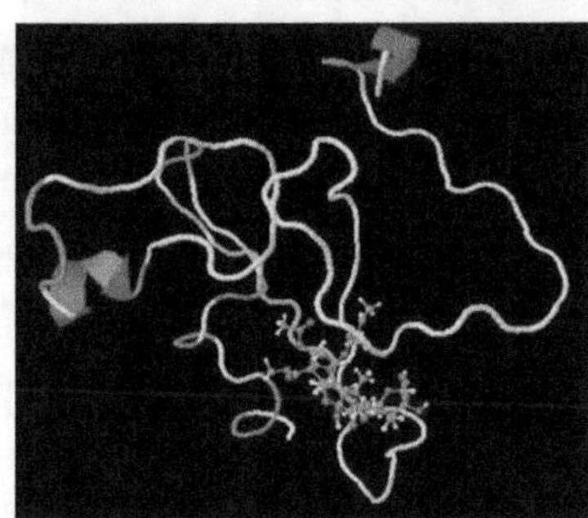

Fig. 47D Interaction with azadirachtin

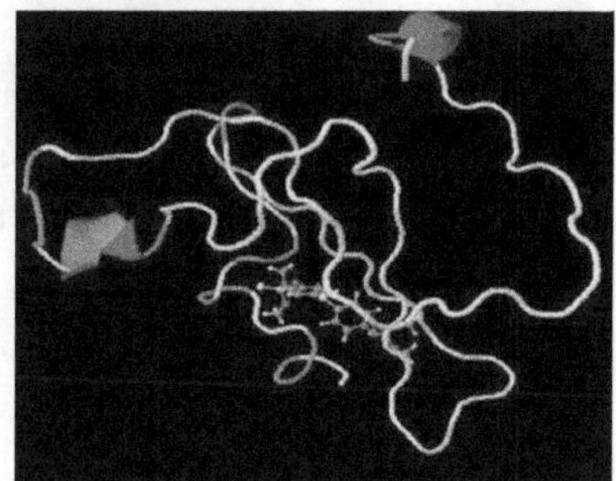

Fig. 47E Interaction with temephos

Fig 47 *In silico* molecular Docking analysis of homology model of *ace1* receptor of *Ae. aegypti* interacted with Stigmasterol (47A); Campesterol (47B); Beta sitosterol (47C); Azadirachtin (47D); Temephos (47E)

6. SUMMARY

1. Mosquitoes are problematic vectors that are dangerous to the living community and spread deadly diseases. The present study aims to isolate and discover the potent larvicidal compounds from the selected seagrasses, *Enhalus acoroides*, *Halophila ovalis*, *Syringodium isoetifolium*, *Halodule uninervis* and *Cymodocea serrulata,* collected from the Mandabam coastal regions, Ramanathanapuram district, Tamil Nadu, India.

2. Mixed stages of larvae were collected from Kodungaiyur, North Chennai, Tamil Nadu and transported to the laboratory followed by segregation and identification based on morphology and movement. The pathogen free cultures of larvae were reared in the laboratory for conducting experiments.

3. The seagrasses were washed, dried, grounded and extracted using three different solvents such as hexane, dichloromethane and ethanol using a rotary evaporator at 50°C.

4. Phytochemical analysis was performed in hexane, Dichloromethane and ethanolic solvent extracts of five different seagrasses to test the presence of aminoacids, protein, tannins, saponins, flavonoids, alkaloids, quinones, glycosides, terpenoids, phenols, steroids and coumarin.

5. Mosquito larvicidal activity was carried out for all the solvent extracts of five different species of seagrasses and the highest mortality rate was recorded in 500 ppm of hexane extracts of *C. serrulata* leaves as 90% and 94% against *C. quinquefasciatus* with the LC_{50} values of 0.022 and 0.129 µg/ml; LC_{90} values of 0.336 and 0.817 µg/ml similarly, against *Ae. aegypti* with the LC_{50} values of 0.006 and 0.030 µg/ml; LC_{90} values of 1.364 and 2.019 µg/ml after 24 and 48 hous, respectively.

6. The impact of epiphytes anchored to *C. serrulata* leaves was studied through larvicidal analysis and it was understood that the efficacy was not due to the epiphytes.

7. The potent crude hexane extract of *C. serrulata* leaves was subjected to bioguided assay fractionation, about 1-20 fractions obtained were combined to eight fractions (A-H) and larvicidal activity carried out on eight fractions and efficiently answered in Fraction E.

8. The TLC pattern of fraction E showed two compounds 'A' and 'B' with the 'R_f' value of 0.4 and 0.33 respectively, and further separated through sub column bio guided fractionation.

9. Among the separated compounds, compound 'A' was found efficient against *C. quinquefasciatus* with the LC_{50} value of of 0.004 and LC_{90} of 0.011 μg/ml compared with controls azadirachtin and temephos.

10. The potent compound 'A' was subjected to characterisation techniques such UV-Vis spectroscopy, FTIR, ^{1}H NMR and GCMS. The UV spectrum showed greater intensity at 255 nm, could be a sterol group and various functional groups from 3430-840 cm^{-1} were observed in FTIR analysis, ^{1}H NMR shows presence of three sterol groups in H-3 at δ 3.52 (multiplet), H-6 at δ 5.36 (triplet) and Individual stigmasterol at H-22 and 23 at δ 5.15 and 5.05 (Multiplet); GC-MS analysis displayed three major sterol groups stigmasterol, campesterol and betasitosterol in the ratio (1:1:1).

11. Histopathological study of *C. quinquefasciatus* was carried out using untreated water, temephos and compound 'A' of *C. serrulata* leaf extracts. From the study, it was found compound 'A' influenced the muscle layer, midgut regions inner cell components and food bolus compared to temephos. whereas, untreated larvae was found normal.

12. The effect of compound 'A' of *C. serrulata* leaves was analysed for non-toxic organisms (predatory fishes) such as *Poecilia reticulata* and *Gambusia affinis* were found non-toxic up to 5 ppm and the SIF/PF was favourable.

13. The binding energy of phytosterols were individually analysed by interaction with *ace1* receptor of *C. quinquefasciatus* and *Ae. aegypti* through *in silico* molecular docking analysis, the binding energy was found as -7.21, -7.29 and -7.63 Kcal/ Mol against *C. quinquefasciatus*. Similarly, the binding energy of 7.20, -7.39, -8.04 Kcal/Mol calculated against *Ae. aegypti* they can be compared to Azadiractin and Temephos.

7. CONCLUSION

Mosquitoes are dreadful vectors since it bites the living communities for blood. In the recent reports of WHO, 2020, about 229 million cases were affected in 2019 and 65% of children below five years are affected by malaria. To restrict mosquito transmission, chemical insecticides are used in breeding places all over the world, however this has a harmful influence on the ecology. Then the studies begin to attract natural substances, and some researchers are also hopeful about marine species. In this regard, seagrasses have been discovered to have therapeutic properties that can be used to suppress mosquito larvae.

In this current research, the hexane extracts of *Cymodoceae serrulata* were tested and analyzed in-depth and have shown the potent larvicidal activity against third instar larvae of *Culex quinquefasciatus* and *Aedes aegypti* which deserted of preliminary phytochemical analysis by Limmermann burchard test that gives the result of presenting the sterol compounds, UV-Vis spectroscopic analysis displayed a peak at 255nm that illustrates sterol groups. In the end, GCMS analysis exhibited three major peaks of phytosterols (stigmasterol, campesterol, and beta-sitosterol) in ratios of 1:1:1 respectively. Therefore, the efficient larvicidal action seen in this study could be related to the possibility of synergism of phytosterols. Furthermore, they can be fully comprehended by experimenting with individual sterols and comparing it to steroidal mixes for the same mosquito strains.

The compound applications of the current study explored that the phytosterols disturbed the midgut regions of larvae in a similar way as temephos and it is non-toxic up to 5 ppm against *Poecilia reticulata* and *Gambusia affinis*. However, these three phytosterols are difficult to separate using conventional chromatographic techniques but the individual phytosterol efficacies was investigated in depth through *in silico* docking research, where the binding energy was discoved to be efficient like temephos and azadirachtin.

It is concluded that phytosterols as synergistic or individual existence, has the ability to perform a significant role in the control of mosquito larvae that are harmless for other organisms and environment. In the future, this can be applied in field trials for the complete examination of the next advanced level of research. So it is recommended that people should be empowered with adequate knowledge on seagrass extract-based insecticides and others to take the necessary actions at the community level to prevent mosquito-borne diseases.